MATH

IS PRECISE, PERIOD,

VS.

MATH

IS PRECISE,
STRINGS ATTACHED

MATH

IS PRECISE, PERIOD,

VS.

MATH

IS PRECISE,

STRINGS ATTACHED

REFLECTIONS OF A MATH TEACHER
ON TEACHING MATHEMATICS

WILLIAM J. ADAMS

Mathematics Department
Pace University

with illustrations by
Ramunė B. Adams

Library of Congress Control Number: 2009905936
ISBN: Hardcover   978-1-4363-3419-8
                         Softcover   978-1-4363-3418-1

This book is available on the web at webpage.pace.edu/wadams

This book was printed in the United States of America.

To order additional copies of this book, contact:
Xlibris Corporation
1-888-795-4274
www.Xlibris.com
Orders@Xlibris.com
40317

"A teacher affects eternity; he can never tell where his influence stops."

Henry Adams
The Education of Henry Adams (Ch. 20)

To most of my Teachers

Acknowledgements

I am indebted to my daughter Ramunė for preparing the illustrations, to my daughter Rasa for helping me to prepare the manuscript, and to the Sabbatical Leave Committee of Dyson College of Pace University for their support.

Contents

Introduction: Current Mathematics Education in Total is a Disaster

despite the many excellent, dedicated teachers who inspire their students to study and learn mathematics.

This book is the outcome of the afore conclusion I reached based on many years of teaching mathematics. It is my hope that by sharing my experience and thinking about teaching math I might be helpful to colleagues, students, and others concerned about mathematics education.

Mathematics education disaster in what sense? No, it's not in the sense that I believe insufficient attention is being paid to number fundamentals. This is a separate issue.

I was recently introduced as a professor of mathematics to a professor whose specialty is the history of architecture. "I'm in a soft social science, but mathematics is solid", he commented to me. This view of mathematics is almost unanimously held. 1+1 is 2, no ifs-ands-or-buts, no ten sides to the story, is often used as an example in support of "mathematics is solid", a dimension which is found by many to be one of its most attractive features. This dimension is the one we are most familiar with from our study of math, but when it comes to math's applications to the study of real-world situations the supposed "solid math" can become as soft as the softest social science. Views on what assumptions should be made in applied studies may vary considerably, and ten sides to the story may well arise.

On this fundamental point current mathematics education is wholly inadequate.

The number is king in the quantitatively oriented world we find ourselves. While statements such as 1+1 is 2 are correct as statements of "pure mathematics", as it is termed, in its application dimension questions about the reliability of the numbers that arise, how relevant they are to the real-world situation being studied, and how they are to be interpreted warrant serious consideration.

On this crucial point current mathematics education fails us.

And then there is the demise of the big IF. Mathematical statements are conditional: if A then B. More and more in current mathematics education we see the playing down of the hypothesis A. The favored attitude is one that is computer oriented. Throw your data into the computer and let it do its thing.

Current mathematics education does us a serious disservice by going further and further in this direction rather than focusing attention on the big IF.

Current mathematics education does not prepare us for life in the 21st century, which requires an understanding of the mathematical modeling perspective, of what mathematics can do and its limitations, and an appreciation of the questions that should be considered to help us distinguish numbers that inform from those that deceive.

If the wizards of Wall Street had a 21st century mathematics education there is a good chance that they would not have put unquestioning faith in their value at risk (VaR) math models and the financial meltdown of 2008-09 would have been avoided, or at least softened. If the nation's decision makers and the public at large were better educated about what questions to give thought to when numbers continually hurled at them are the basis for decision making, they would be less vulnerable to accepting faulty numbers and all of us would be less at risk to the consequences of bad decision making.

These matters are what deeply concern me and are the basis for my rationale for what at first reading might strike you as a strange choice of title for this book. I chose it because I believe that it best expresses the nature of my concern about what I believe to be the disastrous state of current mathematics education.

Consider the following question: Do the following statements reflect math insights, myths, in a sense a bit of both, or nonsense?

- A presentation supported by figures is more credible than one that is not.

- Numbers are neutral. They are not affected by cultural, economic and political differences that influence people and therein lies their strength. $1 + 1 = 2$, for example, is universal for all peoples, countries, and political systems.

- Numbers give weight to a view through the sense of *precision* they communicate, thereby advancing it to a plane which commands recognition, respect, and acceptance.

- Mathematical proof is the most reliable means for objectively establishing truth.

- The *precision* of mathematical methods guarantees unassailable conclusions which serve as pillars of stability and strength in a world besieged by foggy thinking, prejudice, and rampant special interests.

- Mathematically derived conclusions are indisputable because they are based on deductive logic, which is untainted by bias and ideology.

- Considering the availability of powerful computers, in applications of mathematics the hypothesis of theorems warrants minor consideration.

What do we want mathematics education to achieve? The objectives are many, but the fundamental one for 21st century mathematics education I submit, is to leave our students with a perspective on mathematics that would enable them to address the *afore* question.

The generally accepted anthem of mathematics is Mathematics is *Precise*, Period. My years of teaching mathematics convinces me that a more appropriate anthem would be Mathematics is *Precise*, Strings Attached. Strings attached are like the small print in a legal document which we tend not to look at too closely, often to our subsequent regret. And so it is with mathematics. In one form or another strings attached are present at all levels of mathematics, and when they are ignored the results promised must seriously be considered open to question.

For this reason I believe that the key to achieving the afore fundamental objective lies in a change in perspective on mathematics from mathematics is *precise,* period, to mathematics is *precise,* strings attached.

Mathematics teaching and textbooks have become so focused on getting the *right* answer by employing math technique (factoring algebraic expressions, solving equations, calculating the derivatives and integrals of functions, and the like) that in danger of being lost in the shuffle is development of perspective on what *right* answer means and conditions that must be satisfied for math technique and technology turned loose to yield the *right* answer. As we have come to appreciate, the growing availability of powerful technology to implement math technique has made it possible for more and more people having less and less understanding of what math technique and technology can do and its limitations to generate more nonsense more quickly than ever before.

Consider the following scenario:

The Austin Company, a producer of high quality electronic home entertainment equipment, has decided to enter the digital tape player market by introducing two models, Ultra (DT-1) and Supreme (DT-2). Their problem is to determine the number of units of each model that should be produced weekly to maximize profit.

The Company's operations research department was asked to study the problem and make recommendations. The OR department began its analysis by collecting data. They viewed the manufacturing process in terms of three phases; construction, assembly, and finishing. The data collected and their analysis led them to introduce the following **assumptions**, which they took as **postulates**.

P1:   In the construction phase each DT-1 unit requires 2 hours of labor and each DT-2 unit requires 3 hours of labor. At most 1,100 hours of construction time are available per week.

P2:   In the assembly phase each DT-1 unit requires 5 hours of labor and each DT-2 unit requires 3 hours of labor. At most 1,400 hours of assembly time are available per week.

P3:   In the finishing phase each DT-1 unit requires 4 hours of labor and each DT-2 unit requires 1 hour of labor. At most 756 hours of finishing time are available per week.

P4:   After taking the cost and revenue factors into consideration the anticipated profit for each DT-1 unit is $150 and the anticipated profit for each DT-2 unit is $120. In order for these unit profit values to be realistic the Company must produce at least 25 DT-1 and 40 DT-2 units per week.

P5:   There is an unlimited market for the DT-1 and DT-2 models.

P: Other factors that may play a role in determining the production schedule that would maximize profit are not being considered. (This may be for various reasons. They may view some factors as negligible to profit maximization. There may be factors whose importance they recognize, but do not know how to handle. And then, possibly, there are factors that are important, but whose significance they have not recognized.)

Their next task was to translate these **postulates** into mathematical form, being careful to include everything stated in them but not go beyond what is being assumed. The OR department began by introducing variables for the quantities it sought to determine; they let x denote the number of DT-1 and y the number of DT-2 units to be made weekly.

The afore conditions led them to the math problem:

Maximize $P(x, y) = 150x + 120y$
subject to:
$x \geq 0, y \geq 0$
$2x + 3y \leq 1100$

$$5x + 3y \le 1400$$
$$4x + y \le 756$$
$$x \ge 25, y \ge 40$$

Application of math technique yields the solution (100, 300) with maximum value 51,000.

When I ask my students if they would implement this result by setting the production schedule to manufacture 100 DT-1 and 300 DT-2 units per week with an anticipated weekly maximum profit of $51,000, no ifs-and-or-buts, those who reply say yes, of course, without thought being given to the strings attached that should be considered. From their perspective there are no strings attached. Application of math technique yielded the *right* answer and that was that. (Those who do not reply, the student silent majority, seem either confused or thoughtful.)

I introduce the idea of strings attached, discuss what must be considered, with a focus on the issue of the realism of the postulates, and then add the following ingredient to the stew.

The Austin Company also hired the Aleksa Company, a consulting operations research firm, to independently study their digital tape player problem and make recommendations. The Aleksa OR group viewed the manufacturing process in terms of two phases: construction (which included assembly) and finishing. The data collected and their analysis led them to introduce the following **assumptions/postulates**.

P1a: In the construction phase each DT-1 unit requires 8 hours of labor and each DT-2 unit requires 5 hours of labor. At most 2,210 hours of construction time are available per week.

P2a: In the finishing phase each DT-1 unit requires 3 hours of labor and each DT-2 unit requires 2 hours of labor. At most 860 hours of finishing time are available per week.

P3a: The anticipated profit for each DT-1 unit is $140 and the anticipated profit for each DT-2 unit is $150. In order for these unit profit values to be realistic the company must produce at least 50 DT-1 and 50 DT-2 units per week.

P4a: There is an unlimited market for the DT-1 and DT-2 models.

P: Other factors that may play a role in determining the production schedule that would maximize profit are not being considered.

The afore conditions led the Aleksa Company to the math problem:

$$\text{Maximize } P(x, y) = 140x + 150y$$

subject to

$$x \geq 0, y \geq 0$$
$$8x + 5y \leq 2210$$
$$3x + 2y \leq 860$$
$$x \geq 50, y \geq 50$$

where x represents the number of DT-1 and y the number of DT-2 units to be made weekly.

Application of math technique yields the solution (50, 355) with maximum value 60,250.

Which of the two solutions, if either, should be implemented? This question prompts me to introduce a spokesman for a widely accepted view. Robert Turner (a.k.a. Bottom-line Bob), a computer "expert" and member of the board of directors of the Austin Company said: "it's a no-brainer; 60,250 is larger than 51,000 and it was obtained by use of the most sophisticated computer technology available. It's *infallible*; go with the Aleksa Company's solution since we want to maximize profit."

Mention sophisticated computer technology, reaffirmed as yielding *infallible* conclusions, then that's enough to carry the day. Implement the Aleksa Company's solution, no question about it. Alas, of course it's far from being a no-brainer; there is a fundamental question concerning the realism of the assumptions made for which this scene provides us a platform for discussion.

I return to the Austin Company's profit maximization scenario in chapter 8.

As another illustration, I consider what I call the man-in-the-park scenario.

One day the following conversation took place between P.M., a practical man, and S.P., a student of probability, who were sitting on a park bench.

P.M. took a die from his pocket and said to S. P.: "Sir, I understand that you are a student of probability; would you help me with a problem?"

S.P.:   "What is your problem?"

P.M.:   "What is the probability of throwing an even numbered face with this die."

S.P.:   "It's ½?"

P.M.:   "How did you get ½?"

S.P.:   "Take the events 1 shows, 2 shows, 3 shows, 4 shows, 5 shows, 6 shows as your basic outcomes for the tossing of your die. There are six basic outcomes, three of which describe an even numbered face showing. Therefore, the probability that an even numbered face will show on a toss of your die is 3/6 which, of course, is ½."

P.M.:   "Are you sure about this?"

S.P.:   "No question about it; 3 divided by 6 is ½. The *precision* of math makes this conclusion *infallible*.

My students (with perhaps a few silent exceptions) would accept this result without question because of their belief in the *precision* of math. After all, 3 divided by 6 is ½; another 'no-brainer'. There is, of course, much more to it than that which hinges on the sense in which math is *precise*.

P.M.:   "I plan to participate in a 'friendly' game with this die tonight. What is the significance of this value for me?"

S.P.:   "If your die is tossed a large number of times, an even numbered face will show about 50% of the time; you can count on it."

I add the following postscript to this situation in an attempt to stimulate further thought.

P.M. participated in a 'friendly' game that evening and bet his money based on his expectation that an even numbered face would show about 50% of the time when his die is tossed a large number of times. His die was tossed 1000 times, but an even numbered face showed only 200 times, yielding the ratio 1/5, which deviates considerably from the predicted ratio close to ½. He found himself $400 poorer and, confused and angry, he went looking for S.P. where he had found him the day before. He found him on the same park bench and demanded an explanation of what had gone wrong and why math had failed him.

What would be an appropriate explanation? I return to the man-in-the-park scenario in chapter 11 (secs. 11.3 and 11.4).

My experience convinces me that the major obstacle to overcoming the difficulty students have with such scenarios is the math mind-set they had acquired from the "standard" mathematics education that now prevails: math is *precise*, its conclusions are *indisputable* (almost always understood to mean in the real-world sense), no ifs-ands-or-buts about it, is the perspective that they take with them from their school years. One of my honors class students who had breezed through technique oriented math in high school and who I had exposed to such scenarios put it to me this way: "I got 74 on your exam because you didn't ask enough about normal math." Normal math, to him, meant technique oriented math, plain and simple.

In terms of his math experience my going beyond math technique, which he had been brought up on and did well with, made it my fault that he was having difficulty with what he considered abnormal math. I remember him and his like-minded classmates as examples of the poisonous effect that the Math is *Precise*, Period, mind-set can have on bright students.

And then there is the question, what do we mean by math *precision* and math *proof*? The thinking of the Uncle George school of thought on these matters, introduced through the scenario of Aunt Alice's birthday celebration (ch. 4, sec. 4.3), provides us with a platform to address this question.

My concern with this dimension of the Math is *Precise*, Period mind-set led to part 1, consisting of raw materials I found useful in addressing it.

A second dimension of the prevailing view of mathematics *precision* is centered on The Lost IF. The Lost IF is, I'm sure, familiar to many, if not most, teachers. A student, much distressed, goes to his teacher claiming that he had promised him a passing grade, to which the teacher replies: "you forgot the IF part of the promise; I promised you a passing grade IF you passed the final exam, which you did not do."

Alas, the Played Down or Lost IF in the applications of mathematics is far too often the prevalent state of affairs in mathematics education.

My concern about the contribution The Lost IF dimension makes to mathematics education/miseducation is the subject of part 2.

A third dimension of the prevailing view of mathematics *precision* is with what I believe appropriate to describe as Number Madness. It wasn't too long ago that if you gave a talk during which you spoke utter nonsense, the chances were good that you would be denounced and hooted off the stage. On the other hand, if you had the talk committed to print and published, it took on a new life, being viewed as sacred, and cited in books claimed to be purveyors of wisdom.

These days the number has supplanted the printed word as a sacred object and the more startling the number, the greater the wisdom that must be behind it is a far too widely accepted view. The development of an antidote for Number Madness requires that sources of number misunderstanding and deception be identified and a warning sign posted: CAUTION, these numbers should not be accepted at face value.

My concern about the contribution Number Madness makes to the Math is *Precise*, Period mind-set is the subject of part 3.

In the final chapter 27, I return to and consider the seven initially stated candidates for math insight, myth, in a sense a bit of both, or nonsense.

W.J.A.

Overview of the Contents

As to the order in which parts and chapters are taken up, a linear order running straight through from the beginning is fine, but not necessary. However, since chapters 1-4 are basic to setting the stage for the rest of part 1, I recommend that they be read before choosing a selection from the rest of part 1.

For details about the nature of the parts and chapters see the Comprehensive View of the Contents.

Introduction: Current Mathematics Education in Total is a Disaster

1 *Math Modeling, Math Precision, and Math Proof*

2   The Lost IF

3   Number Madness

Comprehensive
View of the Contents

Introduction: Current Mathematics Education in Total is a Disaster

1 Math Modeling, Math *Precision*, and Math *Proof*

In Part 1 I should like to share with you my thoughts on a number of situations that lend themselves to developing and illustrating the mathematical modeling way of thinking, the nature of mathematical *precision*, and *proof* as they are generally understood/misunderstood.

With perhaps a few exceptions, the technical math gateway to the scenarios discussed here ranges from very basic to basic. Math technique is usually not the obstacle those being introduced to math modeling encounter. The obstacle is departure from deep-rooted beliefs to a very different outlook that poses the challenge, as it is with every body of knowledge.

1 Prelude to Math Modeling

This chapter focuses on a (relatively) simple situation for introducing the math modeling point of view, one which I term Jules Warner's model for determining the financial cost of smoking. An alternative model is introduced to provide a setting for discussion of a different approach to this problem.

2   Math Modeling to Determine the Travel Time for a Vacation Trip

This is a step up in sophistication from the cost of smoking models. This situation provides a setting that permits further exploration of the meaning of *right* and *wrong* in math model building, how these issues are settled, and the idea of refining a math model to make it more realistic.

3   More on Truth/Realism, Validity, and Math Modeling Relationships

Venn diagrams to the rescue, plus . . . .

4   Mathematics is *Precise*! How So?

The sense in which this claim is understood/misunderstood is examined through consideration of four scenarios.

5   A Challenge to the Uncle George School of Thought About the Nature of Math *Proof*

Uncle George's understanding/misunderstanding that the truth of the theorems of Euclidean Geometry is established by *precise* math reasoning (ch. 4; sec. 4.3) is one that is almost universally held. The discussion presented here is intended as a challenge to the school of thought he belongs to.

Consideration is given to making the standard of math *proof* that the Uncle George school of thought accepts as *precise* more airtight.

6   Math Modeling in the Algebra Landscape

Functions, graphs of functions, equations and their math-world vs. real-world nature are considered.

7   Math Modeling via Systems of Linear Equations

The focus is on two problems that lead to math models defined by systems of linear equations.

8   Math Modeling to Achieve Profit Maximization

Center stage is taken by the Austin Company's profit maximization problem employed as an illustration in the Introduction. The Austin Company's problem is an important step up in sophistication from the Gaja Company's problem considered in chapter 4 (sec. 4.2).

9   Math Modeling for Judgments Expressed in Quantitative Form

Two problems involving judgments expressed in quantitative form about items or people to be selected which is, in some sense best, are examined.

10 Index Number Modeling

The Consumer Price Index, developed as a measure of the behavior of inflation, is perhaps the most well-known example of an index number whose behavior has far reaching consequences for us all. The difficulties inherent in math modeling in general are illuminated through consideration of this important special case.

11 Probability Modeling

The man-in-the-park scenario presented in the Introduction is further discussed (sec. 11.3). Consideration of probability models for "simple" random processes (such as die tossing) provide us with an opportunity to study up-close the issue of realistic vs. unrealistic math model building. Subjective probabilities and normal curves from a probability modeling view are discussed.

12 Math Modeling for Conflict Situations

Model building concerning parties with conflicting interests is discussed.

13 Math Modeling in Terms of Matrices

Consideration of a math model formulated in terms of matrix equations.

14 Tales from the Land of Differential Equation Modeling

An interesting dimension of math modeling is to be seen by employing concepts and methods studied in calculus to study decay processes.

15 Math Modeling for the Study of the Motions of Celestial Bodies

Selected valid conclusions of Newton's model for the motions of celestial bodies are considered from the point of view of their realism. The aftermath of the discovery that some of these predictions are not realistic is discussed.

16 Math Modeling for the Study of the Structure of Space

Euclidean Geometry's status as a "perfect" description of space is examined. The development of non-Euclidean Geometry that led to its dethroning from this status to that of a math model of space is discussed.

17 Which Approach to a Problem is Preferable?

The statistics landscape provides us with an appropriate setting for consideration of the problem of deciding which of two math model approaches is preferable for addressing a problem.

18 Is Social Security on the Brink of Bankruptcy?

Oft-repeated "wisdom" has it that the answer is yes, but beware the assumptions.

19 Comments on Selected Food for Thought Questions: 1

I thought this might help put into better perspective what I have in mind.

2 The Lost IF

20 The Strings Attached to a Math Structure: A Big Deal?

Very much so. If the gap between the strings attached to a math structure and the situation to which it is being applied is "wide", the conclusion drawn from it in the situation under study is compromised. Ignoring this issue provides comfort and support for the propagation of the Math is *Precise*, Period, mindset.

3 Number Madness

Number madness is another dimension of the Math is *Precise*, Period mind-set. "If you want *precision*, go to the numbers; that's as good as it gets", is a commonly heard refrain.

Alas, on closer inspection we often find that the seemingly solid Numerical Rock of Gibraltar facing us has troubling fissures.

In Part 3 I should like to share with you my thoughts on what math education can do to help those who are brought into contact with numbers (which includes just about all of us) to become more sensitive to these fissures, be able to identify them, and develop perspective on the issue of interpreting numbers.

21 Reliability: Are These Numbers Trustworthy?

Number slinging has become a commonplace practice, but often the reliability of the numbers being so freely bandied about is seriously open to question.

22 Relevance: Which One is the *Right* Number Trail?

"Let's be *precise* about what we're talking about, get the numbers", is the advice often heard echoing throughout the doings of modern life. The key question that warrants serious thought is, which numbers are relevant to the undertaking being planned?

23 Methods for Obtaining the Numbers

The question of reliability is always our companion.

24   Interpretation: What Do the Numbers Tell Us?

Smith says that the numbers mean this; Jones claims that the numbers mean that. Who, if either, is close to reality's mark?

25   Why Do "Bad" Numbers (and Models we Might Add) Matter?

In a word, consequences.

26   Comments on Selected Food for Thought Questions: 3

I thought this might be helpful.

27 Return to the Question of Math Insight, Myth, in a Sense a Bit of Both, or Nonsense

1

Math Modeling, Math *Precision*, and Math *Proof*

1

Prelude to Math Modeling

1.1   Preface

How to introduce math modeling in a simple, interesting way that is not overshadowed by math technique, is the question that faces teachers when we want to introduce the math modeling concept. The vehicle I develop here was in part suggested by an experience.

For many years an artist friend of mine declined to act on the health warnings of his doctor about the consequences of smoking. But one day the doctor inquired as to whether he had any idea about how much it was costing him, and proceeded to note a few figures. My friend gave up smoking immediately. Add to this background an article by Hubert Herring [1] that caught my attention, and we have the birth of Jules Warner and Janet Wright, the stars of my math model vehicle.

1.2   Jules Warner's Math Model for the Cost of Smoking

Jules Warner considered the case of a teenager who takes up smoking at fifteen and persists with the practice for 50 years. The price of a pack of cigarettes will vary over 50 years, but considering what the current price range of a pack is, suppose $4.00 is taken as an estimate of the average price of a pack over the next 50 years, Jules reflected. A new smoker may begin modestly with a few cigarettes a day, and then work up to a pack, 2 packs, and perhaps eventually 3 packs or more a day. Take a pack and a half as an estimate for the average amount smoked per day, over 50 years, Jules thought it reasonable to assume based on the data he had obtained. This yields a cost figure of 1.5 packs per day, times $4.00 per pack, times 365 days, equals $2,190 per year, on average. Multiplying this figure by 50 yields $109,500 for a period of 50 years.

If our smoker gets married and has children he would probably give serious thought to obtaining life insurance, Jules considered. Insurance rates differ for smokers and non-smokers, and what might cost a smoker $1000 a year might only be on the order of $700 a year for a non-smoker, for a $300 difference, Jules' data suggested to him. If the insurance is held over 30 years, say, the additional cost for a smoker comes to $9000.

And then there are cleaning bills. If our smoker is not happy about his clothes, furniture, carpets, and draperies reeking of the smell of smoke and decides to do something about it, the extra cleaning bills might come to the order of $400 per year. Assume 30 years for this too, Jules thought, and the cost comes to $12,000.

And then there is the matter of clean teeth. If our smoker is sensitive about yellow teeth, he might require at least one extra cleaning a year at, Jules thought it reasonable to assume, $90 a cleaning. If 30 years is taken for this too, we have $2,700 as the total cost for teeth cleaning.

An important step in developing the math modeling way of thinking is to recast the background stated in discursive form into the more formal postulate—theorem form.

Jules Warner's Postulates and Theorems

Postulates

P1.   A person takes up smoking at fifteen and continues the practice for 50 years.

P2.   The estimated average cost of a pack of cigarettes over the 50 year period is $4.00.

P3.   The estimated average amount smoked over 50 years is a pack and a half a day.

P4.   The additional cost of life insurance for a smoker is $300 a year. This applies over a period of 30 years.

P5.   The additional cost of cleaning clothes, furniture coverings, drapes, carpets, etc. for a smoker is, on average, $400 a year over 30 years.

P6.   The additional cost of teeth cleaning for a smoker is, on average, $90 a year over 30 years.

Theorems

T1.   The estimated cost of cigarettes over 50 years is $109,500.

T2.   The estimated additional cost of life insurance over 30 years is $9,000.

T3.   The estimated additional cost of cleaning clothes, furniture coverings, drapes, carpets, etc. over 30 years is $12,000.

T4.   The estimated additional cost of teeth cleaning over 30 years is $2,700.

T5.   The estimated total cost of smoking over 50 years is, rounded off, $133,000.

It all looks so *precise*, and in a deductive-logical sense, which defines the meaning of mathematical sense, it is; if you grant Jules' postulates P1 through P6 as a starting point, you must also grant his theorems T1 through T5 as being valid consequences of them. T1 follows from P1, P2, and P3; T2 follows from P4; T3 follows from P5; T4 follows from P6. The mathematical tool in each of these cases is our good old friend multiplication. T5 follows from T1 through T4 by another familiar old friend, addition.

I find this to be a good opportunity to make the point that humble as these familiar old friends might strike us, they are as successful in getting the job done in the sense of *validity* as the most sophisticated cutting edge mathematical tools around today.

I reaffirm this when more sophisticated math tools emerge, as with the corner point theorem in linear program modeling in chapter 8, for example.

1.3   Additional Strings Attached

The story thus far presents us with an appropriate place to introduce the key question that underlies the fundamental distinction to be made between a conclusion's (mathematical) *validity* and its (real-world) *truth*. In theory it's a simple distinction, but I find that in math modeling situations many students find it a difficult distinction to make because of what I term red herrings that distract them from seeing the distinction.

My principle task, I have found, is to help them to identify the red herrings and set them aside.

There are additional strings attached to this story that we should not overlook. These are concerned with factors Jules chose not to address, and this, I feel, should be explicitly noted as a postulate.

P.   Other factors that may play a role in the cost of smoking are not being considered.

This may be because Jules does not know how to come to grips with them or because he views them as insignificant, or possibly a combination of both. It is not current practice to explicitly list P with the postulates of a math model, but I believe that we would be wise not to allow it to fall into the out-of-sight, out-of-mind black hole.

What other factors are not being considered in this model? In an article which prompted the refinement developed here, Hubert Herring [1] notes F1 through F4, to which I would add F5, F6, and F7.

F1.   In a study conducted by the National Bureau of Economic Research it was found that smokers earn 4 to 8 percent less than non-smokers. As Herring points out, "this is a tricky statistic."

F2.   Some 200,000 fires a year are started by smoking materials, which makes smokers more vulnerable to catastrophic loss.

F3.   Expensive puffs. Some smokers have been known to take connecting flights so they could have a cigarette break at an airport, rather than taking a smoke-free cross-country flight.

F4.   Smokers generally eat less than non-smokers, which translates to a saving on food.

F5.   Smokers are generally more vulnerable to a number of health related problems, which translate to a cost. One dimension of this is seen in the difference in life insurance costs for smokers and non-smokers.

F6.   Smokers' family members are generally more vulnerable to health related problems caused by secondary smoke, which translates to a cost.

F7.   Inflation. The $133,000 figure is in terms of today's dollars. Jules did not attempt to adjust for inflation over 50 years.

Disagreement

I believe that an important point to be made at this juncture is that Jules Warner's model is not the sole perspective on the cost of smoking. This is the role of what I term Janet Wright's model (sec. 1.4), which paves the way for discussion of who's *right* and who's *wrong* (sec. 1.5).

On examining the postulates of Jules' model, Janet Wright, based on data she had obtained, was prompted to take issue with the realism of P2, P3, P5, and P6. Janet's data and study led her to introduce another math model for the cost of smoking by replacing P2, P3, P5, and P6 by P2a, P3a, P5a, and P6a, and introducing postulate P7 on health cost.

1.4   Janet Wright's Math Model for the Cost of Smoking

Postulates

P1.   A person takes up smoking at fifteen and continues the practice for 50 years.

P2a.   The estimated average cost of a pack of cigarettes over the 50 year period is $6.00.

P3a.   The estimated average amount smoked over 50 years is two packs a day.

P4.   The additional cost of life insurance for a smoker is $300 a year. This applies over a period of 30 years.

P5a.   The additional cost of cleaning of clothes, furniture coverings, drapes, carpets, etc. for a smoker is, on average, $600 a year over 30 years.

P6a.   The additional cost of teeth cleaning for a smoker is, on average, $150 a year over 30 years.

P7.   The additional cost of health maintenance for a smoker is, on average, $600 a year over 50 years.

Here too, implicit in the formulation of Janet's postulates is the assumption that other factors that may play a role in the cost of smoking are not being considered.

P.   Other factors that may play a role in the cost of smoking are not being considered.

Theorems

T1.   The estimated cost of cigarettes over 50 years is $219,000.

T2.   The estimated additional cost of life insurance over 30 years is $9,000.

T3.   The estimated additional cost of cleaning clothes, furniture coverings, drapes, carpets, etc. over 30 years is $18,000.

T4.   The estimated additional cost of teeth cleaning over 30 years is $4,500.

T5.   The estimated additional cost of health maintenance over 50 years is $30,000.

T6.   The estimated total cost of smoking over 50 years is, rounded off, $281,000.

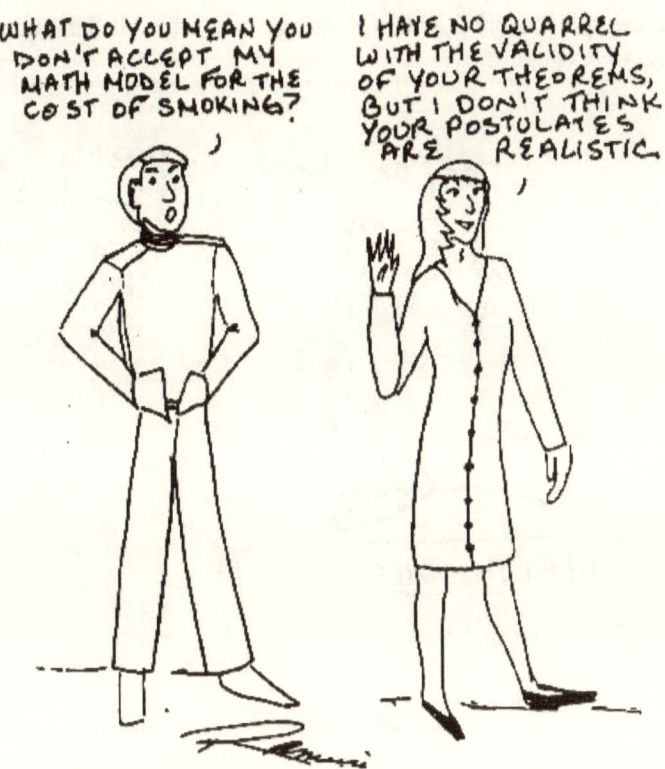

1.5   Who's *Right*: Jules or Janet?

Jules: The total cost of smoking over 50 years is **$133,000. Janet:** The total cost of smoking over 50 years is **$281,000.** Who's *right*? Could both be *wrong*? What does *right* mean? What does *wrong* mean?

This, I have found, is a tough distinction for students to handle. It involves distinguishing senses of *right* and *wrong* that they had not been called on to make in their earlier math studies. For them math conclusions were always *right* if the technique used to obtain them was correctly used, and that was as far as it went.

The difficult point for students to handle, at least initially, is that both Jules and Janet are *right* in the sense of validity, one or the other or perhaps both are *wrong* in the sense of realism, and whether or not they are *wrong* in the sense of realism is irrelevant to the issue of their validity. The giant red herring is to provide information about a valid conclusion not being realistic and raise the question about its not being valid after all. Far too often the answer given is yes.

1.6   Food for Thought Questions

As we appreciate, to help students to get a grip on issues under discussion in mathematics questions for thought are a must. The Food for Thought Questions presented throughout this book are examples of the sort of questioning I have found helpful to them to get a grip on the issues discussed in the Introduction.

Selected food for thought questions are discussed in chapter 19.

1. **Arnold Jacob's Model:** Arnold Jacob set up a math model for the cost of smoking based on the following postulates:

Jacob's Postulates

P1.   A person takes up smoking at fifteen and continues the practice for fifty years.

P2.   The estimated average cost of a pack of cigarettes over the 50 year period is $5.00.

P3.   The estimated average amount smoked over 50 years is 2.5 packs per day.

P4.   The additional cost of life insurance for a smoker is $400 a year. This applies over a period of 30 years.

P5.   The additional cost of health maintenance for a smoker is, on average, $1000 a year over 50 years.

(a)   What theorems can be deduced from Jacob's postulates concerning the cost of cigarettes, the additional cost of life insurance, the cost of health maintenance, and the total cost of smoking?

(b)   Jacob's theorem on the total cost of smoking disagrees with the theorems obtained by Warner and Wright in their models. Does this establish that Warner and Wright's theorems are not valid? Explain. (Red Herring Alert.)

(c)   Jacob's postulates do not take into account the additional cost for smokers of cleaning of clothes, furniture coverings, etc., and teeth, whereas Warner and Wright's models do. Does this mean that Warner and Wright's conclusions on the total cost of smoking are valid, whereas Jacob's conclusion is not valid? Explain. (Red Herring Alert.)

(d) Does the fact that Warner and Wright's models take into account the additional cost for smokers of cleaning of clothes, furniture coverings, etc., and teeth make their models more realistic for the cost of smoking than Jacob's model? Explain.

(e) "Cigarettes Up to $7 a Pack With New Tax," stated the headline of a *New York Times* article (July 1, 2002; B1). Does this result mean that the purported valid conclusions obtained in Jacob's, Warner's, and Wright's models are not valid after all? Explain. (Red Herring Alert.)

1.7   References of Interest?

1. Hubert Herring, "Where There's Smoke, There's Outlay," *The New York Times*, April 27, 1997.

2. William J. Adams, *Algebra with Applications: Technique and THOUGHT*, Revised Edition (Philadelphia: Xlibris, 2009), ch. 5. Available on the web at webpage.pace.edu/wadams.

3. William J. Adams, *Finite Mathematics, Models, and Structure*, Revised Edition (Philadelphia: Xlibris, 2009), ch. 2. Available on the web at webpage.pace.edu/wadams.

4. William J. Adams, *Think First, Apply MATH, Think Further*, (Philadelphia: Xlibris, 2005), ch. 9. Available on the web at webpage. pace.edu/wadams.

5. William J. Adams, *Slippery Math in Public Affairs: Price Tag and Defense* (New York: Marcel Dekker, Inc., 2002), ch. 8.

2

Math Modeling to Determine the Travel Time for a Vacation Trip

2.1   Preface

For many years my family and I have taken vacations in Kennebunkport, Maine. The idea of developing a math modeling situation involving a concern about travel time had its origins in this experience. The stars of the show are Andy, his friend Henry (counterparts of Jules and Janet in the cost of smoking stories), and Andy's sister Rasa. The situations developed in this setting provide us with an opportunity to address the issue of realism of a math model.

2.2   Andy Plans a Vacation Trip

Recently, Andy was engaged in planning a car trip from home in Brooklyn, New York, to the popular vacation town of Kennebunkport, Maine, in mid-August. His problem was to set up a math model for the trip that would enable him to predict the total time required for the journey.

The setting of any such problem presents numerous features and characteristics, many of which are irrelevant or unessential to the focus of the problem. In developing his math model, Andy had to sort this out and decide on which features were fundamental and which were negligible. This required discretion and judgment, the most controversial aspect of the math model development process; one person's essential might be another's irrelevancy.

Andy examined a map and laid out a route. Based on data provided by friends who had recently made the trip, he made assumptions about departure time, weather conditions, the traffic flow to be expected along various points, speeds that would be possible, and the number of rest stops to be made and their duration. Such considerations led him to a math model consisting of a line segment 330 miles long joining points representing Brooklyn and Kennebunkport, and the problem of determining how long it would take an object moving at an average speed of 55 miles per hour to cover this distance.

Andy's math model, it is important to point out, is an idealized, abstract rendering of the real situation involving a trip from Brooklyn to Kennebunkport. It is intended to capture the main features involved in taking such a trip and reflects these features as he sees them and the assumptions that he was led to make. As in the case of Jules vs. Janet concerning the cost of smoking, it is possible that someone else planning such a trip would see things in another light and compose a very different math model.

In my experience this is a difficult point for students to get a firm grip on. Considering that their experience with math had computation and technique as its main focus this is to be expected. I find the analogy of an artist sketching

Andy's portrait a helpful one to make. The artist's sketch is an attempt to capture Andy's main characteristics. It is not intended to be a photographic likeness; if this were so it would be best to use a camera.

Just as another artist might sketch a different portrait of Andy, another mathematical artist might develop a different math portrait/model for Andy's vacation trip time. This point opens the door for consideration of Henry's math portrait/model in section 2.3 along with a number of important questions.

By employing division, we obtain the valid conclusion that an object moving along the idealized path of Andy's model at an average speed of 55 miles per hour would take $330 \div 55 = 6$ hours to make the journey.

In summary, Andy's math model portrait consists of the following:

Andy's Math Model

Andy's Postulates

P1:   A line segment of length 330 miles joining points representing Brooklyn and Kennebunkport is taken as an idealized representation for the actual path of the trip.

P2:   A point moving at an average speed of 55 miles per hour from the point representing Brooklyn to the point representing Kennebunkport along the line segment is taken as an idealized representation of the car trip itself.

P: Factors other than those considered by Andy that may play a role in the time needed to take the trip are insignificant and may be treated as negligible.

What I earlier termed postulate P, recognizing that there are other factors involved in the outcome of what is being studied, but they are not being considered, is here too being explicitly stated. As noted in chapter 1, I believe it important to do this because it makes clearer that there are numerous factors that are candidates for consideration in the math model building process, that a judgment call has to be made about which are to be considered and which are to be set aside, and that there is no one-and-only way to make this judgment call.

Andy's Theorem

T1:   The time required to make the trip is 6 hours.

> **proof:**
> $330 \div 55 = 6$

Just as with Jules' cost of smoking model, this provides a reaffirmation of the point that as humble as good old simple division is as a tool for obtaining a *valid* conclusion, it is as successful as getting the job done in the sense of *validity* as the most sophisticated mathematical tools available to us.

Interpretation

The trip, following Andy's suggested route, will take approximately 6 hours.

2.3   Another Opinion: Henry's Model

Henry sketched a different math portrait/model for Andy's trip time.

For Jules Warner there was Janet Wright to introduce an alternative model for the cost of smoking. For Andy the role of Janet is played by his friend Henry.

Rather than first heading East along the Belt Parkway and then North to Kennebunkport Andy's friend Henry suggested an alternate route which involved first taking the Belt Parkway North through Manhattan and then going East to Kennebunkport (see Figure 2.1).

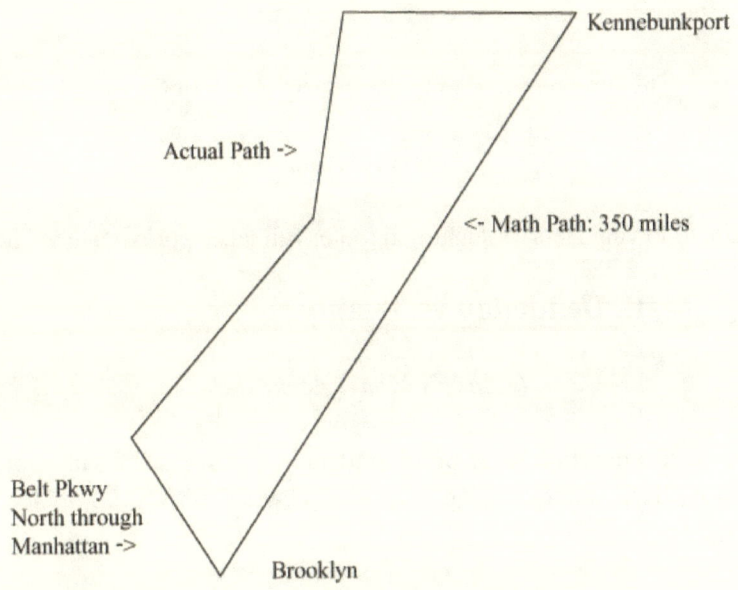

Figure 2.1

Henry's Model

Henry's Postulates:

P1a: A line segment of length 350 miles joining points representing Brooklyn and Kennebunkport is taken as an idealized representation for the actual path of the trip.

P2a: A point moving at an average speed of 50 miles per hour from the point representing Brooklyn to the point representing Kennebunkport is taken as an idealized representation of the car trip itself.

P: Factors other than these considered by Henry that may play a role in the time needed to take the trip are insignificant and may be treated as negligible.

Henry's Theorem:

T1a: The time required to make the trip is 7 hours.

Interpretation:

The trip, following Henry's suggested route, will take approximately 7 hours.

2.4   Math Deduction vs. Reality

Is Andy's Model Accurate?

In reply to this question some of my students say yes, that's what we proved. (Others sit back, some looking somewhat bewildered, waiting to see what direction I would take.)

This sets the stage for making the all-important observation that the answer to this question is to be found by appealing to the real world, not the validity of mathematical conclusions. Undertake the journey on the route Andy laid out, note the time required, and compare it with the projected time of 6 hours from Andy's theorem. If there is, in some sense, a small discrepancy between the actual and projected times, this would establish that Andy's theorem is realistic in this case and, by reflection, be evidence in support of the realism of his math model portrait of the journey. If there is a "large" discrepancy between the actual and projected times, this would establish that Andy's theorem is not realistic in this case and, reflecting back, lead us to conclude that his model is not a realistic portrait of the journey at hand.

I also reaffirm, which I do as often as suitable occasions arise, that the status of Andy's theorem as a valid conclusion derived from his postulates is not at stake here. Division, yielding $330 \div 55 = 6$, establishes the validity of Andy's conclusion and confers on it the status of theorem, no matter what reality's judgment might be.

Is Henry's Model Accurate?

The observations concerning the accuracy of Andy's Model apply to Henry's Model as well.

2.5   Andy's Model vs. Henry's Model

My intent in introducing the Andy—Henry travel models is to provide us with a vehicle for returning to the questions introduced with the Jules—Janet cost of smoking problem, adding to them an additional level of sophistication. This device for extending the scope of this kind of questioning is employed throughout the book. The situations change but the same questions arise, and if I succeed in helping my students to recognize this I believe that my major objective in introducing them to the math modeling way of thinking will have been successful.

Food for Thought Questions

1. Prove Henry's theorem T1a. Does your "proof" of Henry's theorem establish that if Andy follows Henry's route the trip time will be approximately 7 hours? Explain. (Red Herring Alert.)

2. How could Andy determine which, if either, of the models before him is realistic? Explain.

3. Is it possible that neither model is realistic? Explain.

4. Is it possible that both models are realistic? Explain.

5. Andy reflected on his two models and asked his cousin Algis for his advice. "Which one should I implement to take my vacation trip?" "It's a no-brainer," Algis replied. "You want to make the trip in the shortest time, so implement the model you formulated."

   Do you agree with Algis? Explain. (Red Herring Alert.)

6.   Andy implemented the model he had formulated and the trip took him 5 hours and 55 minutes. "That proves it," said Algis to Andy. "Proves what?" asked Andy. "It proves that Henry's conclusion T1a on the travel time is not valid."

   (a)   Do you agree with Algis? Explain. (Red Herring Alert.)

   (b)   If you do not agree with Algis, what does the 5 hours and 55 minutes travel time prove, if anything? Explain.

2.6   Rasa's Vacation Trip

Enter Rasa, whose role is to provide us with a situation in which Andy's model is unrealistic. Questions: What is the cause of it being so? What can be done in the way of repairs?

Andy's sister Rasa was planning to take a brief vacation trip to Kennebunkport during Labor Day weekend. Andy's model worked well for him and she decided to follow the route it prescribed, expecting the journey to take around 6 hours.

Rasa took the trip Labor Day weekend as planned, but it took her 7 hours. This actual trip time differs considerably from the projected 6 hour trip time of Andy's theorem so that something clearly went wrong; but what? "Your theorem stinks," Rasa shouted at her brother in a somewhat agitated manner. "It's *wrong*, it's not valid," she continued. Of course, Rasa's experience proved Andy's theorem *wrong* in terms of reality, but not validity. The distinction, I reaffirm to my students, is a fundamental one. In the confrontation between what actually happened and what the theorem projects to happen, what actually happened—reality—wins. Andy's theorem is a false statement as a description of the travel time to Kennebunkport on a Labor Day weekend, but it remains a theorem. It is still an inescapable consequence of Andy's model—more specifically, his postulates ($330 \div 55$ is still 6) and this is what makes a theorem a theorem.

Rasa's experience, I again reaffirm, sends us a signal that Andy's postulates are not realistic for travel to Kennebunkport on a Labor Day weekend. In reexamining Andy's postulates, we find that they do not realistically take into account unusually heavy traffic delays, characteristic of holiday weekends, around the tollgates of the Whitestone Bridge. Further examination of Rasa's actual trip shows that this is where she had the difficulty.

2.7   Rasa's Refinement of Andy's Model

Rasa undertook to modify Andy's model to make it more realistic for travel to Kennebunkport on a Labor Day weekend. She reviewed Andy's **assumptions** and took into account data on traffic delays around the tollgates of the Whitestone Bridge on holiday weekends. This led her to following refinement of Andy's model.

Rasa's Model

Rasa's Postulates

P1:   A line segment of length 330 miles joining points representing Brooklyn and Kennebunkport is taken as an idealized representation for the actual path of the trip.

P2:   A point moving at an average speed of 49 miles per hour from the point representing Brooklyn to the point representing Kennebunkport along the line segment is taken as an idealized representation of the car trip itself.

P: Factors other than those considered by Andy and Rasa that may play a role in the time needed to take the trip are insignificant and may be treated as negligible.

Rasa's Theorem

T1:   The time required to make the trip is 6 hours and 44 minutes.

Rasa's theorem is in close agreement with her experience, which is evidence in favor of her math model being a realistic portrait of travel from Brooklyn to Kennebunkport on Labor Day weekends under current conditions.

This scenario serves as a simple example of the need to revise a math model in light of the actual doings of the real world.

2.8   NINO vs. RIRO

Useful Follow-Up Observations

Whether Rasa's model will serve as a suitable portrait in the future depends on the extent to which "current conditions" are maintained. Andy's model and Rasa's modified version of it took into account traffic delays due to road repairs that were in progress. When these repairs are completed, current conditions will have changed in a significant way and a suitable modification of both Andy's and Rasa's models would be in order to keep them current.

One way to get a grip on the question of a math model's realism is through its theorems. The other way is through its **assumptions, formulated as postulates**. Are the assumptions realistic? If the answer is *yes*, the assumptions are realistic, and this is a correct assessment, then the theorems will be realistic as well. The RIRO principle operates; if realistic input, then realistic output in terms of theorems. If the answer is *no* to at least some of the assumptions, and this is a correct assessment, then we cannot be sure about the realism of the theorems, I again reaffirm. Some might be right on target, whereas others might be considerably off reality's mark. The NINO principle operates; if nonsense in, there is a good chance of nonsense out, even though some theorems might be realistic.

Rasa was caught by the NINO machine. She took over Andy's model without carefully examining its assumptions and paid the price in terms of an unrealistic theorem about travel time.

None of these considerations, I once again reaffirm, have any affect on the status of Rasa's theorem as a theorem.

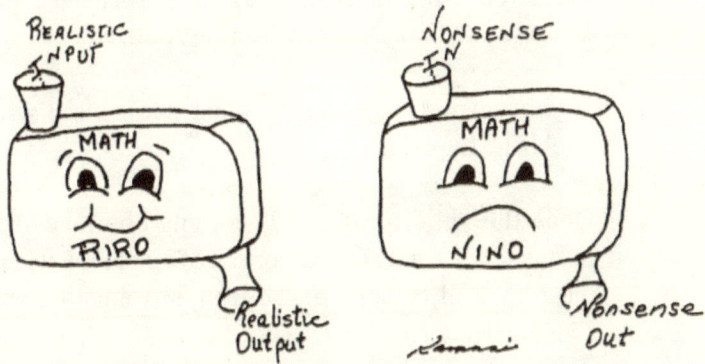

2.9   Food for Thought Questions

7. Rasa's cousin Asta was planning to take a vacation trip from home in Woodhaven, Queens, to Kennebunkport, Maine, at the end of July. Based on data she collected, Asta formulated the following postulates:

Asta's Postulates

P1.   A line segment of length 300 miles joining points representing Woodhaven and Kennebunkport is to be taken as an idealized representation for the actual path of the trip.

P2.   An object moving at an average speed of 60 miles per hour from the point representing Woodhaven to the point representing Kennebunkport along the line segment is to be taken as an idealized representation of the car trip itself.

P.   Factors not considered in the formulation of P1 and P2 have a negligible impact on the vacation trip.

(a)   What theorem can be deduced from Asta's postulates concerning the time it would take to make the trip?

(b)   Asta took the trip at the end of July as planned, but the journey took her six hours. Does the discrepancy between the projected trip time obtained from her postulates and the time it took her to make the trip establish that her conclusion is not valid? Explain. (Red Herring Alert.)

(c)   If your answer to (b) is no, what does this discrepancy establish? Explain.

(d)   Does the validity of the Theorems obtained from the postulates of a math model portrait of a situation under study mean that all or none of the theorems must be realistic? Explain. (Red Herring Alert).

8.   Ann was planning to take a vacation trip from home in Brooklyn, New York, to Kennebunkport at the end of July with a major stop at Putnam, Connecticut for the annual picnic held there. Based on data she collected, Ann formulated the following postulates:

Ann's Postulates

P1.   A line segment of length 250 miles joining points representing Brooklyn and Putnam and one of length 110 miles joining points representing Putnam and Kennebunkport is to be taken as an idealized representation for the actual path of the trip.

P2.   An object moving at an average speed of 50 miles an hour from the point representing Brooklyn to the point representing Putnam is to be taken as an idealized representation of the first leg of the car trip. For the second leg it's an object moving at an average speed of 55 miles per hour.

P3.   Time spent at the Putnam picnic: 2 hours.

P.   Factors not considered in the formulation of P1, P2, and P3 have a negligible impact on the vacation trip.

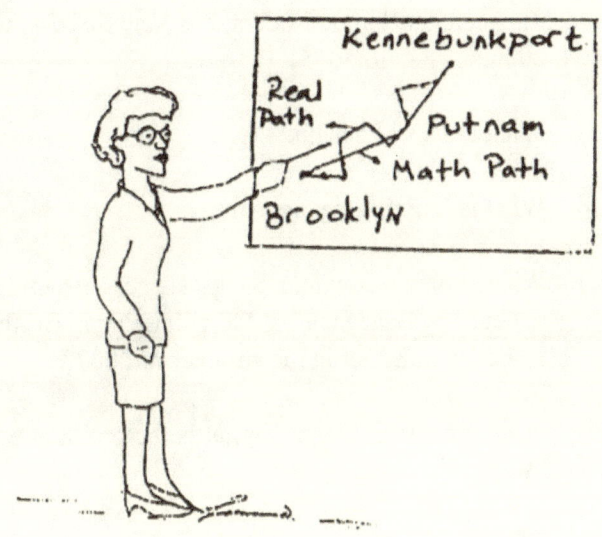

(a) What theorem can be deduced from Ann's postulates concerning the time it would take to make the (i) first leg of the trip?, (ii) the second leg?, (iii) the trip in total from Brooklyn to Kennebunkport?

(b) In making the trip it took Ann 5 hours and 35 minutes to cover the first leg, 1 hour and 50 minutes to cover the second leg, and 9 hours and ten minutes to make the entire trip from Brooklyn to Kennebunkport. Does this establish that the conclusion you obtained in answer to (a), part (i), is not valid, but that the result you obtained in answer to (a), Part (iii), is valid? Explain. (Red Herring Alert.)

(c) If your answer to either or both parts of (b) is no, then what do the given data establish? Explain.

(d) Does the validity of the theorems obtained from the postulates of a math model portrait of a situation under study mean that all or none of the theorems must be realistic? Explain. (Red Herring Alert.)

9. Read "Risk Management" by Joe Nocera, *The New York Times Magazine*, January 04, 2009; 24-33, 46,50, 51.

    (a)   What is VaR? Explain.

    (b)   What is VaR'S great appeal?

    (c)   What explains Goldman Sacks' success in avoiding the pain suffered by Bear Stearns, Merrill, Lehman Brothers and the rest of Wall Street in the summer of 2007?

    (d)   What are Black Swans? What is the problem with a Black Swan?

    (e)   Give an example of a Black Swan that might arise in Andy's vacation trip that is not accounted for in his model.

    (f)   As noted, Rasa's initial vacation trip model did not account for traffic delays around the tollgates of the Whitestone Bridge and this is the principal cause of her model's not being realistic.

            What factors contributed to the VaR models being unrealistic?

    (g)   Does the failure of the VaR models render their conclusions invalid? Explain (Red Herring Alert).

    (h)   (i)   Is the failure of the VaR models a failure of math or of management? Explain.

            (ii)   What about in Rasa's situation? Was the failure of her initial model a failure of math or of Rasa? Explain.

Concerning math models viewed as guaranteeing wealth enhancement on the eve of our great financial meltdown (2008), Warren Buffett, regarded by many as the preeminent financial genius of our time, commented:

"Beware of Geeks Bearing Formulas"

Conversation with Charlie Rose
Oct 1, 2008

Translation: Beware the Assumptions. Again it comes down to the assumptions.

2.10   References of Interest?

1. William J. Adams, *Algebra with Applications: Technique and THOUGHT*, Revised Edition (Philadelphia: Xlibris, 2009), ch. 5. Available on the web at webpage.pace.edu/wadams.

2. William J. Adams, *Finite Mathematics, Models, and Structure*, Revised Edition (Philadelphia: Xlibris, 2009), ch. 2. Available on the web at webpage.pace.edu/wadams.

3. William J. Adams, *Think First, Apply MATH, Think Further*, (Philadelphia: Xlibris, 2005), ch. 10. Available on the web at webpage.pace.edu/wadams.

4. William J. Adams, *Slippery Math in Public Affairs: Price Tag and Defense*, (New York: Marcel Dekker, Inc., 2002), ch. 9.

3

More on Truth/Realism, Validity, Math Methods, and Math Modeling Relationships

3.1   Preface

The mini math proofs presented in the previous chapters on the cost of smoking and Andy's vacation trip have the advantage of using familiar tools from arithmetic, but still, confusion about the senses in which the terms true and valid are applicable does not give way easily.

Approaching truth/validity relationships through use of a visual aid is a useful tool in the war against equating the meanings of true and valid, and Venn diagrams are just the right tool. The approach I employ is the content of sec. 3.2.

3.2   Truth/Realism vs. Validity

Deductive Reasoning, Validity, and Truth

To simply illustrate the nature of a valid conclusion consider the two statements

  1.   All x's are y's
  2.   All y's are z's

and the statement

3.   All x's are z's.

Suppose we take statements (1) and (2) as a starting point for the purpose of seeing what conclusions follow as a logical consequence. In general, a collection of statements set down for such a purpose is called an **hypothesis**. Each statement in an hypothesis is referred to, variously, as an **assumption, premise, postulate,** or **axiom,** depending on context.

The **proof** or **argument** which establishes that a purported conclusion does indeed follow as an inescapable consequence of the postulates is said to be a **valid proof** or **valid argument** and the conclusion of a valid proof is said to be **valid with respect to the postulates,** or **hypothesis** made up of the postulates. Valid conclusions of an hypothesis are called **theorems**.

These basic definitions require flushing out if students are to get a secure hold on them. To begin this process return to the mini-system consisting of (1), (2), and (3).

Postulate P1:	All x's are y's.
Postulate P2:	All y's are z's.
Conclusion C1:	All x's are z's.

P1 forces the class of x's within the class of y's, with P2 forcing the class of y's within the still more inclusive class of z's. It follows as an inescapable consequence, I point out, that the x's, like it or not, are forced within the z's, which is the content of C1. Conclusion C1 is valid with respect to postulates P1 and P2; it is a theorem in this system and may be upgraded from C1 to T1, for theorem.

Visual Aid

Picture these relationships by means of diagrams of the following sort. Represent each class by points inside a closed curve as shown in Figures 3.1(a) and 3.1(b). The diagrams help students to get a better grip on the assumed relationships between the x's, y's, and z's.

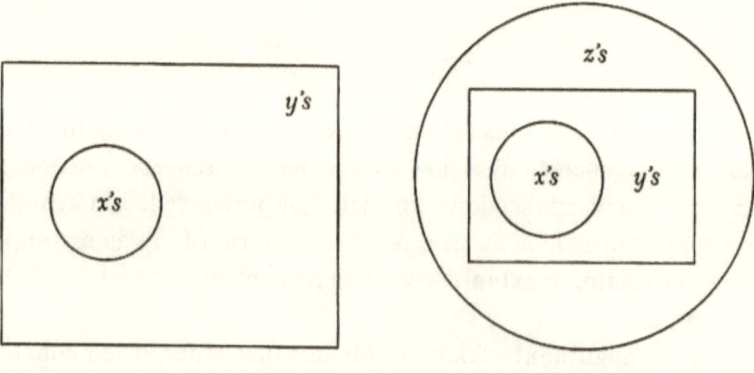

Figure 3.1(a)   *Figure 3.1(b)*

The x's, y's, and z's, I note, are "abstract" entities in the fullest sense of the term and it is important to further note that T1 is a valid consequence of P1 and P2, irrespective of the nature of the x's, y's and z's. The validity of T1 from P1 and P2 is a structural condition, irrespective of the content assigned to the x's, y's, and z's. It's somewhat like having a person in the flesh—this is, the real person—who we may dress up in many ways—business suit, beach wear, evening attire, what have you. The person's appearance changes, sometimes radically, but it is fundamentally the same person. So it is with valid conclusions. We may color-in x, y, and z in many ways, but doing so changes the appearance of P1, P2, and T1, but not the validity of T1 on the basis of P1 and P2.

Suppose, to illustrate, that we add color to x, y, and z as follows. Let:

> x = apple,   y = people,   z = animal

This gives us the hypothesis:

> P1a: All apples are people
> P2a: All people are animals

and the valid conclusion:

> T1a: All apples are animals.

Right vs. *Wrong*

But T1a is *wrong*, many students would say. Well, yes and no; it is crucial that we be very careful, I emphasize, to distinguish the sense in which T1a is *wrong* from the sense in which it is *correct*. T1a is *correct* in the sense of being valid with respect to P1a and P2a; it is *wrong*, that is, false, as a statement about the relationship between apples and animals. (Rasa, we saw in sec. 2.6, was a victim of this misunderstanding.) Adding the coloring x = apple, y = people, z = animal to the scene does not change anything about the structure of the argument which determines validity, but it does introduce a truth/falsity dimension into the scene which, of course, complicates it.

Obvious, you might say. In an immediate sense, yes. But getting it to carry over to more general situations is another challenge. A system made up of postulates and theorems is called, I point out, a **postulate system**. The process of obtaining valid conclusions from the postulates of the system is called **deduction** or **deductive reasoning**. A postulate system M obtained from an **abstract postulate system, APS**, by assigning representations to its abstract terms (x, y, z or equivalents) is called a **model** of the APS. Validity relationships are maintained in passing from an abstract postulate system APS (such as P1, P2, and T1) to any model of the APS (such as M_1 consisting of P1a, P2a, T1a).

If we let

$$x = \text{dog}, \qquad y = \text{mammal}, \qquad z = \text{animal}$$

we obtain model M_2 with postulates

P1b: All dogs are mammals
P2b: All mammals are animals

and theorem

T1b: All dogs are animals.

If we let

$$x = \text{apple}, \qquad y = \text{cat}, \qquad z = \text{fruit}$$

we obtain model M_3 with postulates

   P1c: All apples are cats
   P2c: All cats are fruit

and theorem

   T1c: All apples are fruit.

The models M_1, M_2, and M_3 illustrate, I note, the following relationships between the validity of a statement and its truth/realism.

1. If T is a theorem in a model M of an APS and the hypothesis of M is true/realistic, that is, its postulates are true/realistic, then T is true/realistic. Valid conclusions deduced from true/realistic postulates are true/realistic.

   This is illustrated by M_2. It makes sense since in obtaining a valid conclusion T of an hypothesis H we do not go beyond H. If H is true/realistic and we do not go beyond it to obtain T, then it is not surprising that T is also true/realistic.

2. If the theorems of a model M of an APS are true/realistic, then we cannot conclude that the hypothesis of H of M is true/realistic. H might be true/realistic, some of the postulates of H might be true/realistic and others false/not-realistic; it might be false/not-realistic in its entirety.

   M_3 illustrates a model with a true/realistic theorem arising from false/not-realistic postulates.

3. If a theorem T of a model M of an APS is false/not-realistic; then some of the postulates of M must be false/not-realistic.

   This third property follows from the fact that if the postulates of M were true/realistic, then we could not obtain from them a false/not-realistic theorem T. M_1 illustrates a model with a false/not-realistic theorem (T1a:

All apples are animals) arising from a system with a false postulate (P1a: All apples are people).

In summary:

---

1. **If the postulates are true/realistic, then the theorems must be true/realistic.**

2. **If the theorems are true/realistic, then the postulates may or may not be true/realistic.**

3. **If a theorem is false/not-realistic, then some of the postulates must be false/not-realistic.**

Invalid Arguments

To successfully make the point about valid arguments it is important to illustrate invalid arguments.

Consider the following structure.

Hypothesis. P1: All x's are y's.

P2: All z's are y's.

Conclusion. C1: All x's are z's.

A diagrammatic representation of P1 and P2 is shown in Figure 3.2. The argument is not valid, or **invalid,**

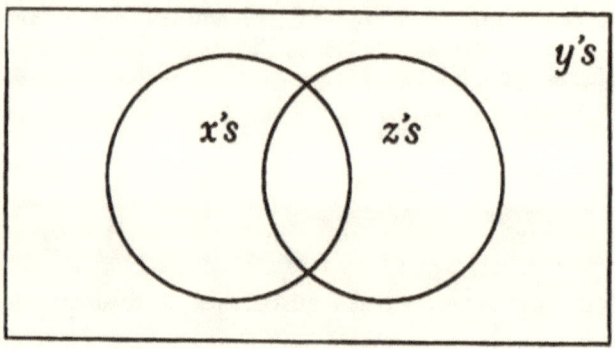

Figure 3.2

because C1 is not forced by the hypothesis. We are forced by P1 to place the class of x's within the class of y's and by P2 to place the class of z's within the class of y's, but this does not force us to place the x's within the z's. It may be that the x's are contained within the z's, but "may be" isn't strong enough for the argument to be valid, I point out.

Food for Thought Questions

Determine the validity of the arguments for the proposed conclusions in 1-6. Explain the basis for your answer.

1. Hypothesis.   P1: All mammals are frogs
                   P2: Joe Warren is a mammal.

    Conclusion.   C1: Joe Warren is a frog

2. Hypothesis.   P1: All oranges are blueberries.
                 P2: All blueberries are fruit.

   Conclusion.   C1: All oranges are fruit.

3. Hypothesis.   P1: Some college students are geniuses.
                 P2: All freshmen are geniuses.

   Conclusions. C1: All freshmen are college students.
                 C2: No freshmen are college students.
                 C3: Some freshmen are college students.
                 C4: Some college students are freshmen.

4. Hypothesis.   P1: All babies are beautiful.
                 P2: Amy is beautiful.

   Conclusion.   C1: Amy is a baby.

5. Hypothesis.   P1: Some x's are y's.
                 P2: All x's are z's.

Conclusions.   C1:   Some z's are y's.

                 C2:   Some z's are not y's.

6.   What is it about the "wisdom" stated in the cartoon that prompted the reply, "What have you guys been drinking?"

As with geometry, diagrams are an invaluable aid for helping us to visualize relationships. To avoid potential misunderstanding of the nature of validity, it is important for us to reaffirm in every suitable situation that the validity of conclusions does not lie with the diagrams, but rather with the relationships themselves. (More on this in ch. 5.)

I find the analogy with chess a useful one to make. Chess can be played without the visual aids of chessboard and physical pieces, and there are those who do so. But for the majority of us mortals chessboard and physical pieces are indispensable visual aids.

3.3   Math Methods

The mini math proofs concerning the cost of smoking and Andy's vacation trip supplemented by the preceding is a useful preface, I find, for more general comments about truth/validity and math modeling.

What do the mini-systems considered in the afore section have to do with the math I have been studying all these years, many students would be thinking.

The answer, of course, is everything: math proof; math methods; math technique are deductive in structure; their application yields valid conclusions from underlying postulates or assumptions which are taken as a starting point. These postulates or assumptions are often not explicitly stated, but they are there. To take an example, being asked to solve the equation x - 3 = 5, we add 3 to both sides to obtain x = 8. Does this simple use of math technique establish that x = 8? Not quite.

The **assumption** made, but not explicitly stated, is that the equation x - 3 = 5 has a solution to begin with. We say solve the equation x - 3 = 5 rather than the more cumbersome, assuming that it has a solution, solve the equation x - 3 = 5. In solving x - 3 = 5 to obtain x = 8 we have not shown that x = 8 is a solution, but that assuming x - 3 = 5 has a solution, the only possibility is 8. We then proceed to replace x by 8 in the equation to verify that it is the solution. This is more than a check against error. It is a necessity, logically speaking; by dong this we are verifying that our assumption that the equation x - 3 = 5 has a solution is correct.

Consideration of this kind of simple situation is helpful in making clear why such **underlying assumptions** are usually not explicitly stated in books on algebra. However, the extreme of never making this point, characteristic of almost all algebra books I know of, does a serious disservice to mathematics education, I would submit.

3.4   Math Modeling Relationships

When postulates involve some aspect of the real world so that it's meaningful to talk about their truth/falsity or, to allow an important shade of gray into the scene, their realism/non-realism, the principle noted in our discussion of the mini-systems explored earlier carry over to the more complex postulate systems explored in subsequent chapters. In these more complex settings the term **model** is replaced by **math model**. The process of **developing a math model** for a situation under study is called **math modeling**.

Pertinent to math modeling are the relationships noted in sec. 3.2 which are at the heart of the math modeling process.

1. If the postulates of a math model M are true/realistic, then the theorems of M must be true/realistic.

2. If the theorems of M are true/realistic, then the postulates of M may or may not be true/realistic.

3. If a theorem of M is false/not-realistic, then some of the postulates of M must be false/not-realistic.

3.5   Shades of Validity

A major problem that we face in our attempts to teach about the meaning of valid/validity/valid conclusions as used in mathematics is that these terms have a number of meanings in our language. In addition to pounding away at the meaning of valid in mathematics through examples and food for thought questions, I believe the following kind of discussion is a useful supplement.

We have employed the term valid/validity in a very specific manner. It is this sense of the term which is to be understood throughout our discussions.

As is the case with many words, valid is used in a number of ways in our everyday language. We might hear, for example, Jim had valid reasons for not following instructions; we must validate our assumptions. Synonyms for valid in our language include sound, cogent, convincing, and telling.

3.6   Food for Thought Questions

7. "A refined radar technique that may settle the current debate over the validity of Einstein's general theory of relativity has been successfully tested." (*The New York Times*, Feb. 28, 1968; 20).

    (a)   Could such a technique be used to settle a question of validity? (Red Herring Alert.)

    (b)   What issue might such a technique help to resolve?

8. "Evidence reported by physicists last fall suggested that the particles [protons], the basic building blocks of matter, had long, but finite lifetimes. But a recent report by researchers who participated in an Ohio study says that the proton lifetimes may be even longer than the billions of years previously estimated. They also say this could mean theories predicting such decay are invalid." (*The New York Times*, Jan. 23, 1983).

    (a)   Would this evidence render the conclusions of such theories of proton decay invalid? (Red Herring Alert.)

    (b)   What effect would this evidence have on theories of proton decay?

9. " . . . theorists [Albert Einstein and Satyendra Nath Bose] calculated [on the basis of quantum theory] that if a certain class of atoms could be chilled to temperatures below any that exist in nature, the atoms would merge with each other to become huge 'superatoms:' . . . The creation of a Bose-Einstein condensate, as this hypothetical superatomic state of matter is called, . . . would not only demonstrate the validity of some outlandish predictions of quantum theory, but would create a form of matter that may never have existed anywhere before." (M. Browne, "Physicists Get Warmer in Search for Weird Matter Close to Absolute Zero," *The New York Times*, Aug. 23, 1994, C1.)

    (a)   Would the creation of a Bose-Einstein condensate demonstrate the validity of some outlandish predictions of quantum theory? Explain. (Red Herring Alert.)

    (b)   What would the creation of a Bose-Einstein condensate demonstrate? Explain.

10. "Precise observations of subatomic particles have uncovered a serious discrepancy with the collection of theories that explain the known particles and forces that shape the universe, scientists at Brookhaven National Laboratory announced." J. Glanz, "Tiniest of Particles Pokes Hole in Physics Theory," *The New York Times*, Feb. 9, 2001; A1.

(a)   Does this establish that some of the conclusions reached from the aforenoted theories are not valid? Answer Yes or No and explain. (Red Herring Alert.)

(b)   If you answered No, then what does the "serious discrepancy" establish? Explain.

11. Consider the following statements; for each one state, with explanation, whether you Agree or Disagree.

(a)   If some of a model's theorems have been confirmed by experimentation/observation, then the model's postulates must be true/realistic.

(b)   If some of a model's postulates are unrealistic, then some of its theorems must be false/un-realistic.

(c)   If a model's postulates are unrealistic, then some of its theorems may be true/realistic.

(d)   If a model's postulates are true/realistic, then some of its theorems may be false/un-realistic.

(e)   Postulates of a model, by their very meaning, are true/realistic statements.

(f)   If a theorem is false/un-realistic, then it cannot be a valid statement.

(g)   If a conclusion obtained from a model is true/realistic, then it must be a theorem of the model.

(h)   A conclusion obtained from a model can be shown to be a theorem by showing that it is true/realistic.

3.7   References of Interest?

1. William J. Adams, *Think First, Apply MATH, Think Further*, (Philadelphia: Xlibris, 2005), ch. 8. Available on the web at webpage. pace.edu/wadams.

2. William J. Adams, *Finite Mathematics, Models, and Structure*, Revised Edition (Philadelphia: Xlibris, 2009), ch. 11. Available on the web at webpage.pace.edu/wadams.

4

Math is *Precise?* How So?

4.1   Preface

One dimension of mathematics that many find attractive is its *precision*: "mathematics is *precise*; its conclusions are true, no argument; one plus one equals two (equated to one apple plus one apple yields two apples), and that's that; no ten sides to the story." This school of thought—which I belonged to in my high school and undergraduate college years (validity was never mentioned.)—obviously never encountered the math modeling dimension of mathematics.

Some, (like myself) eventually do encounter this dimension and appreciate its nature and significance while most, I'm sorry to conclude on the basis of my teaching experience, find it incomprehensible. Their thinking, having been so locked into the math is *precise*, period, mind-set, cannot handle this dimension. As teachers of mathematics, our only hope of breaching the barrier established by this mind-set is through discussion, examples, food for thought questions, repeated by discussion, examples, food for thought questions, repeated by . . . .

In the following I invite you to consider four scenarios which I believe to be a challenging blend of red herrings that are intended to camouflage the distinction between real world truth/realism and mathematical validity, and penetrate the camouflage.

4.2   The Gaja Company's Profit Maximization Problem

The Gaja Company makes shoes. The Company plans to introduce two new styles, for the time being designated by A-18 and A-21, and management wants to know how to set its monthly production schedule to maximize profit. Two consulting firms were hired to study this problem and make recommendations. Each consulting firm formulated a math model, designated by M1 and M2, respectively, for this problem. A theorem obtained from the postulates of M1, interpreted, states that to maximize profit 50,000 A-18 pairs and 35,000 A-21 pairs should be produced monthly for an anticipated monthly profit of $230,000. A theorem from the postulates of M2, interpreted, states that to maximize profit 40,000 A-18 and 50,000 A-21 pairs should be produced monthly for an anticipated profit of $200,000.

As a member of the Board of Directors of the Gaja Company you have been asked for your thoughts on which model should be implemented.

(a) Would you recommend that M1 be implemented because the anticipated profit from M1, $230,000 per month, exceeds the anticipated profit from M2, $200,000 per month? Explain. (Red Herring Alert.)

b) If your answer to (a) is no, what would you recommend?

4.3   Aunt Alice's Birthday Party

While on your way to your Aunt Alice's birthday celebration, her hundredth rumor has it, the world seemed to be in conspiracy against you. It was snowing, traffic was bumper to bumper, and then someone in an old Oldsmobile sideswiped your new Buick and took off. You finally arrive at your aunt's celebration, but it takes you forty minutes to find parking. Needless-to-say, you're not in very good humor, and on entering Aunt Alice's house you find yourself in a social group being "educated" by Uncle George.

Uncle George thinks he knows everything about everything and on this occasion his subject is mathematics, specifically geometry. "Since their *truth* was established by the *precise* mathematical reasoning for which the ancient Greek mathematicians are justly famous, the *truth* of the theorems of Euclidean Geometry is beyond question," bellows Uncle George in his most authoritative sounding tone. Usually, your attitude toward Uncle George is one of toleration, but this time you're ready to tackle bear.

What reply would you give to Uncle George's profound observation?

4.4   Mathematical Methods Have the Advantage of Certitude?

The following point of view was expressed at an economics seminar. Would you agree?

"Mathematical methods in economics have the advantage of *certitude*. No qualified person can resist the *truth* of a mathematical conclusion properly communicated. The job of communication may be difficult if the solution is complex, but when communication is competent, agreement is inevitable. If anyone doubts a solution, he can recalculate the equations and check the steps in the derivation. Then he must either demonstrate that there has been an error or acknowledge the *truth* of the solution."

4.5   Math Will Rock Your World

"Math Will Rock Your World" by Stephen Baker, *Business Week*, Jan. 23, 2006 (pp. 54-62) discusses the role of mathematics—math modeling in particular—in transforming the modern world. It would be difficult to find a more richly textured paeon to mathematics.

As a sample to give you an idea of this texture I quote one section, "How Math Transforms Industries" (p. 57).

Mathematicians have long enjoyed celebrity status in Silicon Valley and on Wall Street. Now they're plying their trade throughout the U.S. economy:

Consulting:

IBM: Big Blue is building math profiles of 50,000 consultants so that computers can pick the perfect team for every assignment. Other tools eventually will be able to track their progress, hour by hour, and rate their performance. Workers will eventually labor in virtual assembly lines.

Food and Beverage:

Enologix: The goal of this California consultancy is to help vintners mimic the chemistry of wines ranked highly by leading critic Robert B.

Parker. It employs algorithms to cull a database of 70,000 vintages and run the analyses. Precise studies of customer data provide blueprints for new products.

Advertising:

Efficient Frontier: The Silicon Valley startup provides mathematical optimization for online ad campaigns. It calculates response rates and return on investment for every advertisement. Broad shift from hunch-based campaigns to mathematical targeting.

Police and Intelligence:

National Security Agency: Mathematicians at nation's top techno-spy agency build algorithms to trawl Internet and phone traffic looking for patterns in speech, subject, and frequency that might point to the next attack. Investigators wade through rivers of data in search of would-be terrorists.

Marketing:

Umbria: Colorado startup assigns numeric values to picks and pans of products that pop up on blogs. Using vector graphics, it confirmed that raunchy Burger King ads online turned off nearly everyone, except for the target audience of young men. Math-based consultancies scour blogs and podcasts for market intelligence.

Media:

Inform: This New York startup turns written articles into bits of geometry and organizes them in a virtual library. It can match the articles to readers' math-based profiles. Automatic systems threaten to supplant editors.

In the closing part of the article the author notes that midcareer managers "still must understand enough about math to question the assumptions behind the numbers."

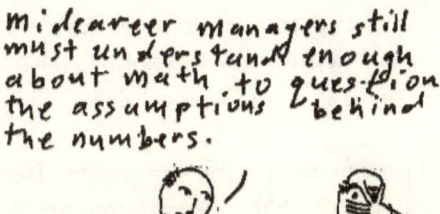

Do you agree or disagree with this assessment? Explain.

4.6   The Afore Scenarios

After introducing the math modeling point of view through the cost of smoking and/or Andy's vacation trip examples I present the afore scenarios to my students, sometimes in the form of exam questions. I would estimate that on average on the order of 50 percent of them get past the red herrings and do a respectable to excellent job in addressing the questions posed. That's the good news; the bad news is that the remaining 50 odd percent don't seem to have a clue as to what it's all about.

The only sensible response, it seems to me, is to persist. Many of the remaining chapters in this part of the book are concerned with the presentation of illustrations that would help us to do just that.

The Gaja Company's Profit Maximization Problem

The red herring is the more attractive $230,000 against the less attractive $200,000 dangled in front of eyes fixed on maximizing profit. Those who go for M1 over M2 because of its more inviting promise have taken the bait. They missed the crucial point that these values do not stand alone in splendid isolation, but are tied to the assumptions/postulates of M1 that underlie $230,000. Are these assumptions/postulates of M1 realistic?, is the crux of the issue.

The same situation arises with Andy's model vs. Henry's model with valid conclusions of 6 hours vs. 7 hours travel time for the vacation trip to Kennebunkport.

Aunt Alice's Birthday Party

If we remove the distraction "for which the ancient Greek mathematicians are justly famous" Uncle George's wisdom reads: "the truth of the theorems [of Euclidean Geometry] was established by mathematical reasoning." One might believe that the theorems of Euclidean Geometry are true, but this is irrelevant to the issue of what math proof establishes, and is a red herring. The question is, does math reasoning establish their truth? Of course not, we would have to reaffirm that math reasoning establishes their validity on the basis of postulates.

The biggest red herring is use of the word truth when it should be validity; one word, but worlds apart in ideas, like peace and war.

Mathematical Methods Have the Advantage of *Certitude*?

It's an attractively presented package, but flawed because here too it is being claimed that math methods establish *truth*.

Math Will Rock Your World

"A key statement in the closing of the article notes that midcareer managers "Still must understand enough about math to question the assumptions behind the numbers." Not so. Knowledge of math technique is useless here. It is intimate knowledge of the enterprise the math is being applied to that will prepare midcareer managers to challenge the realism of the **assumptions**. Math methods applied to unrealistic assumptions cannot be expected to yield golden truths. Garbage in, garbage out—as the saying goes."

—William J. Adams
Professor of Mathematics
Pace University
New York

Readers Write,
Business Week,
Feb. 13, 2006

4.7   Word Power: Watch Your Language!

Careless use of language can be troublesome in any exchange of ideas but as the afore discussions have, I hope, made clear, sensitivity to words is crucial in discussions of mathematical modeling. *Valid* and *true* cannot be used interchangeably without running the risk of complete misunderstanding. *Right* and *wrong* are powerful words, as are *precise, correct, proof, infallible, accurate,* and *certitude.*

Such words evoke strong images which, depending on the context, may or may not be correct. My advice to my students is that when you encounter such words ask yourself, in what sense? Throughout this book I use italic print for such words to help us to remind ourselves to ask, in what sense.

4.8   Food for Thought Questions

1. ZKB Electronics puts out two kinds of personal computers, model ZKB-47 and model ZKB-82. The management of ZKB called in the Aleksa consulting firm to determine how many units should be made

daily to maximize profit. The consulting firm set up a math model for the electronics company's production problem and, by applying math methods, reached the conclusion that 300 ZKB-47 units and 250 ZKB-82 units should be made daily to maximize profit. Before implementing this conclusion, the management of ZKB put the following questions to the director of the consulting firm.

(a)   Does use of math methods guarantee that profit will be maximized when 300 ZKB-47 units and 250 ZKB-82 units are made daily and sold? (Red Herring Alert.)

(b)   What is your basis for recommending that we implement your conclusion?

How would you reply to these questions?

2.   A team of international trade experts headed by Janet Valdez developed a math model for foreign trade between a group of countries in Central and South America. One of the theorems of this model holds that the gross domestic product of each of the participating countries would rise by at least 5% per year for ten years if all tariffs between the countries were eliminated. Before implementing this theorem by eliminating tariffs a commission made up of the trade ministers of the countries involved undertook to review the Valdez model.
     What should be the focus of the review? How so?

5

A Challenge to the Uncle George School of Thought About the Nature of Math *Proof*

5.1   Preface

Deductive *proofs* were first encountered by Uncle George and many of us in a high school course in geometry characterized by the use of diagrams to guide us to the conclusion being established. Diagrams took on a life of their own beyond serving as a guide and the orderly sequence of logical inferences wherein each statement made is justified by a postulate or previously established theorem is broken by appealing to visual evidence to justify statements made in developing the *proof*.

5.2   Dubious Deductions

The following is a well known theorem of Euclidean Geometry, but let us look carefully at the proof that is usually presented (at least in George's day).

Theorem. The base angles of an isosceles triangle are equal. (Today we would say congruent.)

Given: Triangle ABC with $\overline{AC} = \overline{BC}$
To prove: Angle A = Angle B.

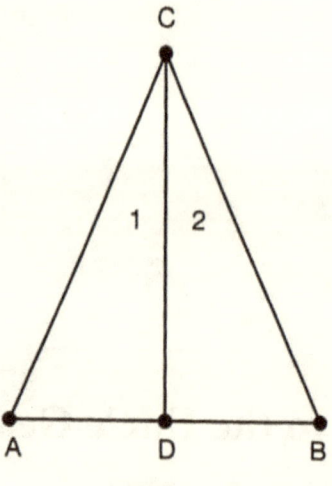

Figure 5.1

Proof.

Statement	Justification
1. Draw the bisector of angle C.	1. Every angle has a bisector.
2. Extend it to meet AB at D.	2. A line may be extended.
3. In triangles ACD and BCD, $\overline{AC} = \overline{BC}$.	3. Hypothesis.
4. Angle 1 = Angle 2.	4. Definition of angle bisector.
5. $\overline{CD} = \overline{CD}$	5. Identity
6. Triangle ACD is congruent to triangle BCD.	6. Side-Angle-Side.
7. Angle A equals angle B.	7. Corresponding parts of congruent triangles are equal.

The proof and diagram are convincing but is it, by itself, a valid argument? Strictly speaking, we would have to say no. A difficulty arises in Step 2. The justification a line may be extended, does not say that a line may be extended to meet another line, AB in this case; the lines might be parallel, and we need something in the system itself that would rule this out.

Let us suppose that we can get around this difficulty and conclude that the bisector of angle C does intersect line AB at D. But where is D? For the proof to hold up we need D to fall between A and B, but there is nothing in the system that says it must. Figure 5.1 is so convincing on this point that our first reaction might be to say that it's "obvious" that D is between

A and *B*; where else could it be? "Obvious" is not the same as deductive proof and postulates and theorems on "betweeness" are needed to support such conclusions.

What is the place of diagrams, George would undoubtly ask. Should they be abandoned. The answer is a resounding NO. Diagrams are invaluable for suggesting ideas and providing us with a sense of what we want to do and what we are obtaining. They are essential allies; but their use should not be equated to formal deductive proof, which must come out of the underlying system itself, we should point out to George.

Purported theorem: There exists a triangle with two right angles.

Proof. Consider two circles that intersect at two points which we shall term *A* and B (see Figure 5.2). Let $\overline{AC}, \overline{AD}$ denote their respective diameters from *A*. Let $\overline{CD}$ meet

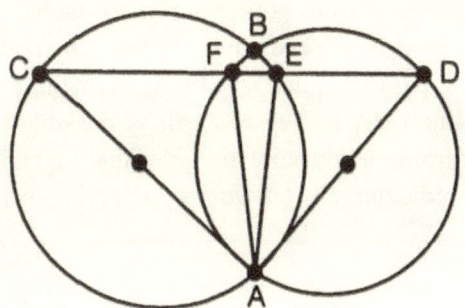

Figure 5.2

the respective circles in *E*, *F*. Thus angle *AEC* is a right angle, since it is inscribed in semicircle *AEC*. Similarly, angle *AFD* is a right angle. Thus triangle *AEF* has two right angles.

    The argument seems airtight, but something is wrong since a triangle, with angle sum 180 degrees, cannot have two right angles. The diagram is compelling, but a very carefully drawn diagram might suggest that $\overline{CD}$ does not pass through *B*, so that *AEF* is not a triangle at all but a line segment. While a very carefully drawn diagram might suggest this, there is nothing in the postulates or theorems of Euclidean Geometry which allows us to argue that $\overline{CD}$ does not pass through *B*.

Food for Thought Questions

Pinpoint the gaps in the following proofs.

1. Playfair's form of the parallel postulate:

    If given a line L and point P not on L, then there is one and only one line which passes through P and is parallel to L (see Figure 5.3).

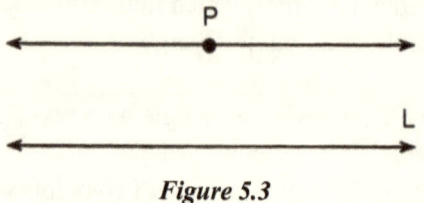

Figure 5.3

Proof

1. Drop a perpendicular from point P to line L (see Figure 5.4). To this perpendicular erect a perpendicular PE from the point P. This second perpendicular is parallel to line L by the theorem that two perpendiculars to the same line are parallel. Since it is possible to drop only one perpendicular from a given point to a given line, and it is possible to erect only one perpendicular to a line from a point lying on it, the parallel line PE is unique.

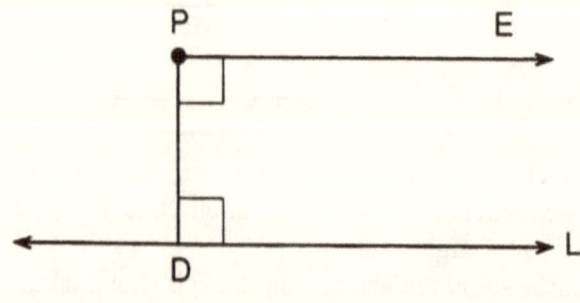

Figure 5.4

2. Return to 1 for a "proof" by the Greek geometer Proclus (410-485). From 1 we have that there exists a line M passing through P parallel to L (see Figure 5.5).

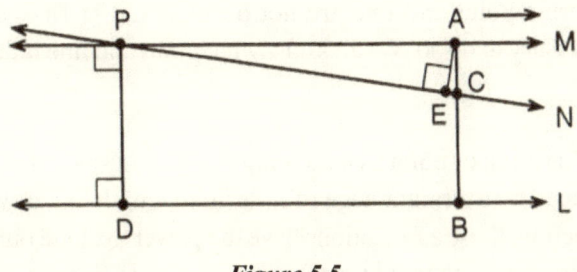

Figure 5.5

The problem is to show that M is unique. Suppose there were another line N through P parallel to L. Then N makes an acute angle with PD that lies on one side or the other of PD. Suppose it's the right side of PD. The part of N to the right of P is thus contained in the region bounded by L, M, and PD.

Let A denote any point of M to the right of P; let AB denote the perpendicular to L at B, and let C denote the point at which AB intersects N. Then $\overline{AB} > \overline{AC}$. Let A recede on M; then $\overline{AC}$ increases without bound since $\overline{AC} \geq \overline{AE}$, the perpendicular from A to line N. Thus $\overline{AB}$, which is at least as large as $\overline{AC}$, increases without bound.

But the distance between two parallel lines must be bounded. Thus we have a contradiction to the result that $\overline{AC}$ increases without bound, which means that our supposition that there is another line N through P parallel to L is untenable.

5.3   Raising the Bar on Math Proof

To help George obtain a better perspective on math proof consider the following mini-postulate-system I_a founded on the following postulates:

P1a: Every line is a collection points which contains at least two points.

P2a: For any two points there is at least one line containing them.

P3a: For any line there is a point not contained by it.

P4a: There is at least one line.

The basic terms point and line are not defined and at first sight it might seem unnecessary to do so. We all know what point and line indicate, George would say,

Yes, most of us think of point as indicating position in space and line as being a path produced by a straight edge of indefinite length but, as we saw in the preceding section, these associations have the power to cloud our minds when it comes to constructing valid proofs.

To get around this difficulty let us use uncharged terms—zog and glob, for example. The abstract postulate system that emerges, call it **Glob Theory**, is based on the following postulates:

$P1$:   Every glob is a collection of zogs which contains at least two zogs.

$P2$:   For any two zogs there is at least one glob containing them.

$P3$:   For any glob there is a zog not contained by it.

$P4$:   There is at least one glob.

What are globs and zogs? George asks. They are undefined. Every postulate system begins with undefined terms, undefined in the sense that no unique characterization in terms of more basic entities is given. There is no way around this. If we were to define glob and zog in terms of mumbo and jumbo, let us say, then mumbo and jumbo would be our basic undefined terms. If we sought to define mumbo and jumbo in terms of more basic elements, then they would be our undefined terms, and on it goes. Postulates $P1$ through $P4$ do not uniquely define zog and glob, but state some relationships between them, we should point out to George.

We now turn to proving some theorems to illustrate proof in this sort of setting.

Possibility: Glob Theory contains at least three zogs.

Idea: Get a glob into play since a glob contains at least two zogs. As a visual aid we might refer "informally" to something like Figure 5.6(a). This gives us two zogs, p and q, and we are well under way.

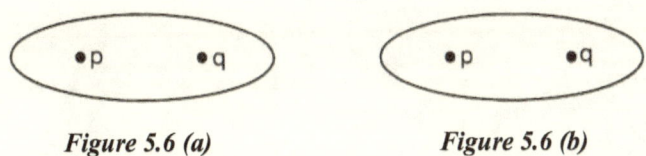

Figure 5.6 (a)   *Figure 5.6 (b)*

We can nail down our third zog r by appealing to $P3$. As a visual aid we have Figure 5.6 (b). These figures are just that, helpful visual aids; they do not comprise the formal proof.

At this point, with these ideas in hand, we are ready to write down the formal proof.

Theorem 1. Glob Theory contains at least three zogs.

Proof

Statement	Justification
1. Let L denote a glob.	1. $P4$
2. L contains at least two zogs, p and q.	2. $P1$
3. There is a zog r not in L.	3. $P3$
4. p, q, and r are distinct zogs.	4. Summary of preceding results.

The observation that the zogs are distinct, with justification, is most important. For convenience we have introduced letters to denote the zogs, but the fact that different letters are being used does not by itself guarantee that the zogs are different, it should be pointed out to George.

What about the globs? How many can we count on in general? Figure 5.6 (b) suggests that there are at least three. Let us see if we can develop a package of ideas to prove this result. A possibility of this sort which is suggested by consideration of special cases is called a **conjecture,** we should inform George.

Conjecture. Glob Theory contains at least three globs.

Idea: we can generate an initial glob, call it L_1, by appealing to $P4$ (see Figure 5.7 (a). We can get other globs into play by introducing zogs and then appealing to $P2$ (see Figure 5.7 (b).

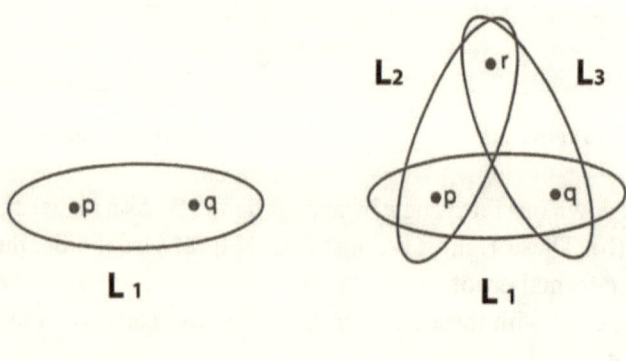

Figure 5.7 (a)   Figure 5.7 (b)

We are now ready to write down a formal proof.

Theorem 2. Glob Theory contains at least three globs.

Proof

Statement	Justification
1. Let L_1 denote a glob.	1. P4
2. L_1 contains at least two zogs, p and q.	2. P1
3. There is a zog r not in L_1.	3. P3
4. There is a glob $L_2 \neq L_1$ which contains zogs p and r.	4. P2; $L_2 \neq L_1$ since L_2 contains r which is not in L_1.

Now we must be very careful. Figure 5.7 (b) is helpful in providing us with a start and initial direction, but it is also potentially misleading. The suggestion conveyed is that we follow up by asserting the existence of glob $L_3 \neq L_1$ which contains r and q and exhibiting globs L_1, L_2, and L_3 to conclude the proof. But how can we be sure that L_3 is not the same as L_2? We can get around this difficulty by introducing L_3 as containing r and q and considering two cases.

5. Case 1. L_3 is not the same as L_1 and L_2, in which case our work is done.

6. Case 2. L_3 is the same as L_2, in which case p, q, and r are contained by the same glob, $L_3 = L_2$, visually illustrated by Figure 5.8 (a).

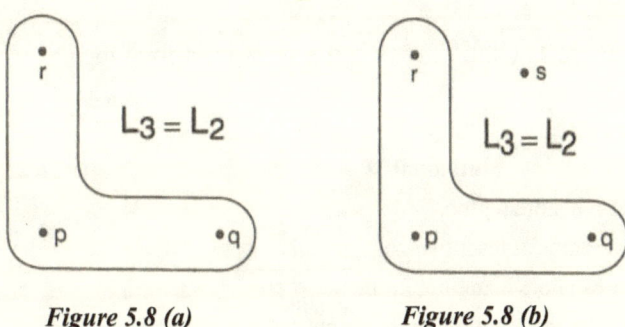

Figure 5.8 (a)   **Figure 5.8 (b)**

Then, by $P3$, there is a zog s not contained by $L_3 = L_2$, visually illustrated by Figure 5.8 (b). By P_2 there is a glob $L_4 \neq L_2$ which contains r and s. $L_4 \neq L_1$ contains r which is not contained by $L_1 \cdot L_1$, L_2 and L_4 are distinct globs.

To this point in its development Glob theory consists of two undefined terms, four postulates, and two theorems. In undertaking to prove other theorems we may now employ theorems 1 and 2 to justify assertions that we make, we should remind George.

5.4   Proof vs. Indication of Proof

As discussed in sec. 5.2, the so-called proof of the theorem the base angles of an isosceles triangle are equal (congruent) based on the diagram in Figure 5.1 serves as the basis for an argument that the afore result is a theorem, but it falls short of being a valid argument. Calling such an argument a proof sows confusion about what a mathematical proof entails.

It would be less confusing and more appropriate to call it an **indication of proof** rather than proof. Most of us would find reading math proofs in their full detail extremely tedious and an indication of proof generally suffices to make the point, which is why this is what is usually done in math books.

5.5   Food for Thought Questions

3. Having established that there are at least three zogs in Glob Theory, we might attempt to climb the mathematical ladder another rung and establish that Glob Theory has at least four zogs.

Is the following argument to prove this result valid? Explain

Conjecture 1: Glob Theory contains at least four zogs.

Proof.

Statement	Justification
1. Let L denote a glob.	1. $P4$
2. L contains at least two zogs, p and q.	2. $P1$
3. Let r denote a zog not on L.	3. $P3$
4. Let K denote a glob containing r.	4. $T2$
5. Let s denote another zog in K.	5. $P1$
6. p, q, r, and s are four zogs in Glob Theory.	6. Summary of preceding steps.

4. Michael Vlasik, an "expert" on Glob Theory, conjectured that this postulate system contains at least six zogs. Is the following argument that Michael developed to prove this conjecture valid? Explain.

Conjecture 2: Glob Theory contains at least six zogs.

Proof.

Statement	Justification
1. Let L, K, and M denote globs.	1. $T2$
2. Let a and b denote zogs in L, c and d zogs in K, and e and f zogs in M.	2. $P1$
3. a, b, c, d, e, and f are six zogs in Glob Theory.	3. Summary of preceding steps.

5. William Schneider, another student of Glob Theory, argued that since the proof given in 3 to prove that Glob Theory contains at least four zogs was invalid, this conjecture is not a theorem of Glob Theory. Would you agree? Explain.

6. Melinda Hu, another student at Glob Theory, expressed the view that Conjectures 1 and 2 are not theorems of this system. Is she justified in this view? Explain.

7. **Neighborhood Theory,** with undefined terms blob and neighborhood, is based on the following three postulates.

*P*1.   There are at least three blobs.

*P*2.   For any two blobs there is at least one neighborhood containing them.

*P*3.   For any two blobs *p* and *q* there is a neighborhood *P* containing *p* and a neighborhood *Q* containing *q* where *P* and *Q* have no blobs in common.

Prove the theorem that there are at least three neighborhoods.

6

Math Modeling in the Algebra Landscape

6.1 Preface

There is a wide gap between functions that arise from real world settings and their mathematical counterparts that are considered in standard algebra courses. Serious attention should be given to identifying the gap and addressing it. This is what this chapter is about.

6.2 Functions: Math-World vs. Real-World

Whatever a student's background might be, it is reasonable to assume that cost, revenue, income and profit are ideas that we can build on. To begin, let us start with cost.

Cost Functions

Consider a firm that is producing a single, uniform commodity. The function

$$c = c(x)$$

which describes the total cost c for the production of x units of output, is called the **cost function of the firm**. Thus, for example, let us **assume** that

$$c(x) = \frac{1}{2}x^2 + 20x + 900$$

is the cost function of a coffee producer, where x is output in tons per day and $c(x)$ is the cost in dollars per ton. Then the cost of producing 10 tons per day is, according to this function

$$c(10) = \frac{1}{2}(10)^2 + 20(10) + 900 = \$1150.$$

Values of $c(x)$ for selected values of x are shown in Table 6.1, and the resulting points are plotted in Figure 6.1. Different scales are required for the x and y axes because of the difference in magnitude between the output values x and cost values $c(x)$. While necessary in this case, these different scales have a distortion effect that we should be aware of and sensitive to. The points plotted in Figure 6.1 almost seem to lie on a line; in fact, they lie on a parabolic arc.

Table 6.1

x	0	4	8	12	16	20
c(x)	900	988	1092	1212	1348	1500

The smallest value of $c(x)$ is 900. Much space would be wasted if we were to indicate the full range of values between 0 and 900 on the y-axis. It is common practice to omit such a range and to indicate this omission by employing the symbol shown on the y-axis in Figure 6.1.

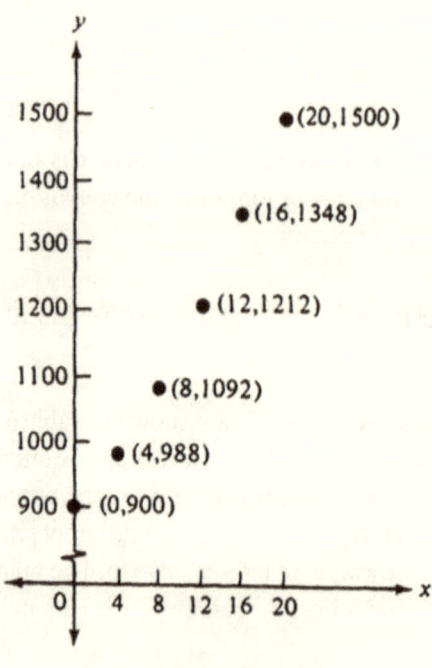

Figure 6.1.

In the standard algebra course we would be considering the afore function $c(x)$ in its own right without reference to an applied setting such as this. Life is easier in the pure math world since the function's domain of definition would be understood to be all real-numbers for which its rule makes math sense, which translates to all real-numbers. Its graph is a parabola, a picture of which can be obtained by plotting some points and connecting them with a smooth curve.

When you have an application setting the only way to deal with such, in my view, is to raise appropriate questions and follow through.

A Fundamental Question

What is the domain of definition of $c(x)$? First, let us note that negative values for x make no sense in this situation. Moreover, output, being expressed in tons of coffee produced per day, is divisible up to a point. Output levels of 1/2 of a ton, 1/5 of a ton, 1/10 of a ton may be feasible from a technological point of view, but what about 1/100 of a ton, 1/1000 of a ton, 1/10,000 of a ton, and the like? At some point, sooner or later, the technological level of the production machinery will not be able to accommodate to such levels. What about irrational numbers such as $\sqrt{3}$, $\sqrt{105}$, $\sqrt{1001}$ and the like? It makes no sense to talk about production at output levels that correspond to such values. And then we should keep in mind that we can only increase output up to a certain level, a level which cannot be identified in a sharp, unequivocal way.

What it comes down to is that the domain of definition of a function that arises from a concrete applied setting is much more complex than one that arises from a mathematical setting in which the restriction is that the domain of definition must make mathematical sense (avoid values that lead to division by 0, for example).

What this example illustrates is the basic fact that reality is a more demanding taskmaster than mathematical considerations by themselves. It is a fact that we should not allow ourselves to forget, for we will have occasion to give attention to a number of reality situations.

An Idealized Mathematical Representation of $c(x)$.

To simplify the problem of dealing with the complexities of reality by making them mathematically manageable we take

$$c(x) = \frac{1}{2}x^2 + 20x + 900$$

defined for $x \geq 0$,

as our mathematical representation, called a **mathematical model,** for the real-world $c(x)$ with its more complex domain of definition. The graph of this mathematical model $c(x)$ is presented in Figure 6.2.

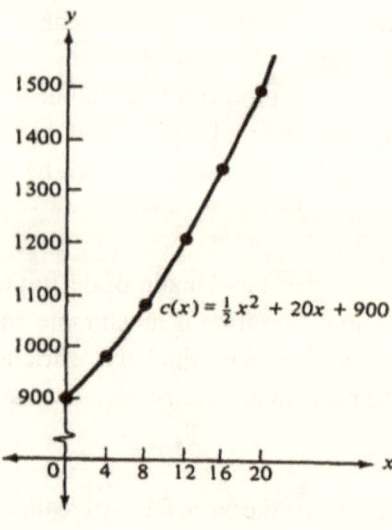

Figure 6.2

Real World Actualities vs. Mathematical Representations

This interplay between real-world actualities and simplified mathematical representations for them, called **mathematical models**, is characteristic of the applications of mathematics.

Cost, Revenue, and Profit Functions

The function

$$R = R(x)$$

which describes total revenue as a function of output, is called the **revenue function of the firm**. If, let us assume,

$$R(x) = 306x - 5x^2$$

is the revenue function of the coffee producer, where x is output in tons per day and $R(x)$ is revenue in dollars per ton, the revenue obtained for an output of 10 tons per day is, according to this function

$$R(10) = 306(10) - 5(10)^2 = 2560 \text{ or } \$2560.$$

The domains of definition of these functions consist of all $x \geq 0$ for which $c(x)$ and $R(x)$ make economic sense.

The profit function $P(x)$ of the firm is the difference between the revenue and cost functions:

$$P(x) = R(x) - c(x)$$

If $c(x) = \frac{1}{2}x^2 + 20x + 900$ and $R(x) = 306x - 5x^2$ are math models of the coffee producer's cost and revenue functions, the math model of the coffee producer's profit function is:

$$P(x) = (306x - 5x^2) - (\frac{1}{2}x^2 + 20x + 900)$$

$$= -\frac{1}{2}x^2 + 286x - 900, \ where \ x > 0$$

With the afore introduction to math modeling in terms of functions my next step is to, in a sense, take a step back by introducing a situation and an **assumption/postulate** which leads to a function, and follow up by employing the kind of analysis as the afore.

The Thomas Company

The Thomas fast-food chain has 60 restaurants in New England, each doing an average of $20,000 worth of business per day. Studies conducted on the impact of opening new restaurants in the region suggest the following **assumption/postulate**:

P1: The average amount of business done by each restaurant in the chain will drop by $200 per day for each new restaurant opened.

Implicit in all such analyses is what I have referred to in earlier discussions as postulate P.

P: Other factors that bear on the situation are not being considered.

This may be the case because other factors that may bear on the situation are viewed as negligible, or it is not understood how to come to grips with them, or a combination of both.

With respect to these conditions, express the total daily average income of all restaurants in the chain as a function of the number of new restaurants opened.

Let x denote the number of new restaurants to be opened. Then $60 + x$ expresses the total number of restaurants that will be in operation, and $20,000 - 200x$ expresses the average daily income of each restaurant. The total daily average income $I(x)$ is the product of the average income of one restaurant, $20,000 - 200x$, and the number of restaurants, $60 + x$. Thus

$$I(x) = (20,000 - 200x)(60 + x).$$

What is the domain of definition of $I(x)$? First of all, $x \geq 0$ since we will either add a certain number of restaurants to the region or not. Negative values for x make no sense because we are not considering selling off existing restaurants. Fraction values, such as $\frac{1}{2}$ and $\frac{3}{4}$, clearly make no sense in this setting. And then there is the question of how large x can realistically be. It is clear that at the very least $I(x)$ must exceed zero if $I(x)$ is to be realistic. This leads us to consider the inequality

$$20,000 - 200x > 0,$$

from which we obtain $x < 100$. Thus, at the very least, x must be less than 100.

In summary, we have the non-negative values $0, 1, 2, \ldots$, perhaps extending as far as 99, as comprising the domain of definition of $I(x)$.

The management of the Thomas Company would like to determine the number of additional restaurants that should be added to the chain to maximize $I(x)$.

To address this question consider the mathematical model counterpart of $I(x)$, which is

$$I(x) = (20,000 - 200x)(60 + x)$$
$$= -200x^2 + 8000x + 1,200,000$$
$$\text{where } 0 \le x \le 99.$$

Needed Background: The quadratic function $f(x) = ax^2 + bx + c$, $a < 0$, takes its maximum value at $x = \dfrac{-b}{2a}$.

Here $a = -200$ and $b = 8000$, so that $\dfrac{-b}{2a} = 20$. Mathematically speaking, we can tell the Thomas Company that $I(x)$ is maximized for $x = 20$.

What Does this Mean to the Thomas Company?

Does it mean that if the Thomas Company were to implement this finding by adding 20 additional restaurants to its chain the total daily average income of all the restaurants in its chain will be maximized?

We cannot say for sure. Mathematically speaking, we have a valid conclusion **with respect to the assumptions made**. How close it is to reality's mark is another issue. This takes us back to the question of **how realistic** are the two starting points of the mathematical analysis, P1 and P.

If the management of the Thomas Company is satisfied that these **assumptions/ postulates** are realistic, then it would be reasonable to implement the mathematical conclusion by adding 20 additional restaurants to its chain.

The following example provides an interesting vehicle for making the point about real world accuracy.

Bacteria Growth

Assuming that each bacterium in a culture consisting of three *E. coli* organisms divides at the end of each second, determine the function that describes the number of bacteria in the culture at the end of t seconds. Find the number of bacteria in the culture at the end of 5 seconds.

At time $t = 0$, the starting point of our study, the culture contains 3 bacteria. At time $t = 1$, each bacterium gives rise to 1 additional bacterium, so that the number of bacteria at time $t = 1$ is

$$3 + 3 = 3(2).$$

At time $t = 2$ each bacterium in the culture gives rise to 1 additional member, so that the number of bacteria in the culture at time $t = 2$ is the number present at time $t = 1$, 3(2), plus the size of the increase, 3(2). Thus at time $t = 2$ there are

$$3(2) + 3(2) = 3(2)[2] = 3(2)^2$$

organisms. At time $t = 3$ each bacterium in the culture gives rise to 1 additional member, so that the number of bacteria in the culture at time $t = 3$ is the number present at time $t = 2$, $3(2)^2$, plus the size of the increase, $3(2)^2$. Thus at time $t = 3$ there are

$$3(2)^2 + 3(2)^2 = 3(2)^2[2] = 3(2)^3$$

bacteria.

More generally, this analysis leads to the function

$$y = 3(2)^t$$

where y is the number of bacteria in the culture at the end of t seconds. The domain of definition consists of values $t = 0, 1, 2$, etc., up to a certain point.

As t becomes larger and larger the accuracy of this growth function diminishes, since it does not reflect such factors as the limited ability of the environment to support life and loss through death.

The predicted number of bacteria in the culture at the end of 5 seconds is

$$y = 3(2)^5 = 96,$$

keeping in mind the afore **assumption**.

6.3   A Market Equilibrium Question

Of interest in the world of economics is the function that expresses the demand D for a commodity (that is, the amount of the commodity absorbed by the underlying market per unit of time) as a function of the unit price, x, of the commodity. The function

$$D = D(x)$$

which expresses demand D as a function of price x, is called a **demand function.**

The domain of definition of $D = D(x)$ consists of all values of x for which the function makes economic sense. These values generally occur in an interval and, for **mathematical convenience**, the set of all values in some suitable interval is adopted as the domain of definition of the corresponding **math model function.**

In many situations, $D = D(x)$ is a linear function. To illustrate, let us **assume** that

$$D = -2x + 150$$

expresses the demand D for coal in thousands of tons per month in a certain region as a function of x, the price of coal in dollars per ton. What is the domain of definition of this function? Negative values of x do not make sense, and x such that D is zero or negative can be discarded as well. To determine x for which D is zero, solve $D = -2x + 150 = 0$ for x.

$$-2x + 150 = 0$$
$$-2x = -150$$
$$-x = 75$$

Thus we take the interval (0, 75) as the domain of definition of the **math model function** corresponding to our demand function. The graph of $D = -2x + 150$ is shown in Figure 6.3

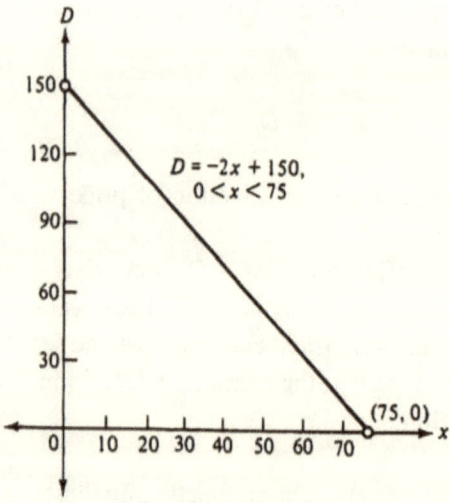

Figure 6.3

Another relationship of interest in economics is the function that expresses the supply S of a commodity (the amount of the commodity that producers make available to the market per unit time) as a function of the unit price, x, of the commodity. The function

$$S = S(x)$$

which expresses supply S as a function of price x, is called a **supply function**. In many situations, $S = S(x)$ is a linear function. For example, let us **assume** that

$$S = 3x$$

expresses the supply of coal in thousands of tons per month as a function of x, the price of coal in dollars per ton. We take all nonnegative values in a suitable interval as the domain of definition of the **math model function** corresponding to our supply function. The graph of $S = 3x$ is shown in Figure 6.4.

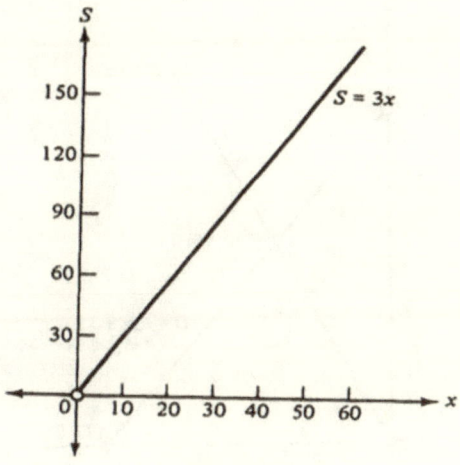

Figure 6.4

If both $D = D(x)$ and $S = S(x)$ pertain to the same market, and x, D, and S are expressed in the same units, then **market equilibrium** corresponds to that point at which the demand and supply curves intersect. The corresponding price and quantity are called the **equilibrium price** and **equilibrium quantity**, respectively.

Under certain conditions that define what economists call **pure competition**, there is a tendency for the price to adjust itself until market equilibrium is attained. The equilibrium price is determined by setting $D(x)$ equal to $S(x)$ and solving for x. From our demand and supply functions for coal, we obtain:

$$3x = -2x + 150$$
$$5x = 150$$
$$x = 30$$

For $x = 30$, $D(30) = S(3) = 90$. Thus the equilibrium price is, according to this analysis, $30 per ton, the equilibrium quantity is 90 thousand tons of coal per month, and the market equilibrium point is $E(30, 90)$.

The situation can be seen in geometric terms by graphing the demand and supply functions on the same coordinate system (see Figure 6.5).

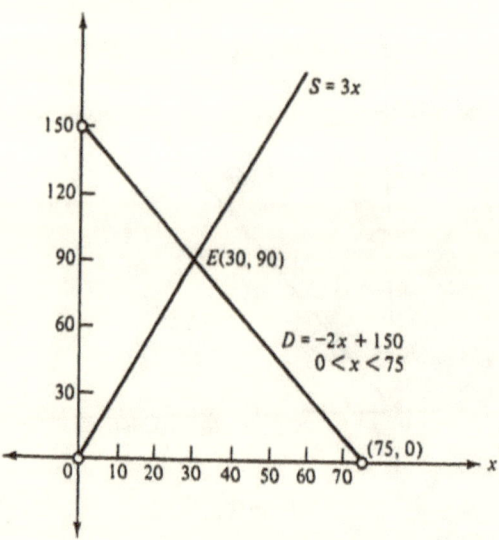

Figure 6.5

Math vs. Reality

This is an issue that comes up whenever an application of mathematics arises. We have the market equilibrium point $E(30, 90)$, mathematically speaking, but its real world accuracy, we should keep in mind, hinges on the real-world accuracy of the demand and supply functions $D = -2x + 150$ and $S = 3x$.

6.4   Food for Thought Questions

1. A taxi fleet uses 100 taxis to service Johnson City. Each taxi brings in an average of $200 per day in fares. If additional taxis are added to the fleet, it is **estimated** (assumed, if you prefer) that the amount in fares brought in by each taxi will drop by $10 per day for each additional taxi added.

   (a) State the function (including its domain of definition) that expresses the total daily average income of the fleet in terms of the number of taxis added to the fleet.

   (b) State the corresponding math model function.

(c)   Determine the value for which the math model function is maximized.

(d)   Will implementing the value found in answer to (c) maximize the total daily average income of the taxi fleet? Explain.

2.   The demand and supply functions for beef in Astin City are **assumed** to be $D = -5x + 50$ and $S = 15x$, where x is the price in dollars per pound, D is demand in thousands of pounds per week, and S is supply in thousands of pounds per week.

(a)   Sketch the graphs of the math model functions of the demand and supply functions on the same coordinate system.

(b)   Determine the market equilibrium point.

(c)   Does this analysis establish the real world accuracy of the market equilibrium point? Explain

3.   The demand and supply functions for orange juice in the summer in Bell City are **assumed** to be $D = -4x + 400$ and $S = 4x$, where x is the price in cents per gallon, D is demand in thousands of gallons per day, and S is supply in thousands of gallons per day.

(a)   Sketch the graphs of the math model functions of the demand and supply functions on the same coordinate system.

(b)   Determine the market equilibrium point.

(c)   Does this analysis establish the real world accuracy of the market equilibrium point? Explain.

4.   Let us **assume** that the cost and revenue functions of a chocolate producer are given by $C = 1000 + 100x$ and $R = 150x$, respectively, where x is output in tons per week, and C and R are in dollars per ton. **Break-even analysis** is concerned with determining the output for which the firm breaks even, that is, the output for which cost equals revenue.

(a)   Sketch the graphs of the math model functions of the cost and revenue functions on the same coordinate system.

(b)   Determine the break-even point.

(c)   Does this analysis establish the real-world accuracy of the break-even point? Explain.

7

Math Modeling via
Systems of Linear Equations

7.1   Preface

The focus is on two problems—one from accounting, the other from traffic network flow—that lead to math models in terms of systems of linear equations. The tableau method, a variation on matrix inversion procedures, is introduced for solving the larger than usual systems of linear equations that arise.

7.2   Service Charge Allocation

Knowledge of accounting is not a prerequisite for this problem.

The Arkin Company, which makes television sets, has three production departments, which we shall denote by P_1, P_2, and P_3, and four service departments—accounting, maintenance, marketing, and purchasing. Each service department's cost must be distributed to the production departments and to other service departments based on their respective usages of the services provided. For each service department listed in the leftmost column of Table 7.1, the fraction of its total cost assigned to the service and production departments of the firm is stated. Thus from row 1 we have that 2 percent of the total cost of accounting is **assumed** to be assigned to maintenance, 10 percent of the total cost of accounting is **assumed** to be assigned to marketing, and so on. Column 1 specifies the fraction of the service departments' costs that is **assigned** to accounting, column 2 specifies the

Table 7.1

SERVICE DEPARTMENT	SERVICE DEPARTMENTS				PRODUCTION DEPARTMENTS			JANUARY OVERHEAD
	Acc.	Main.	Mar.	Pur.	P_1	P_2	P_3	
Accounting	0	0.02	0.10	0.10	0.24	0.26	0.28	$20,000
Maintenance	0.10	0	0.20	0.10	0.20	0.20	0.20	$18,000
Marketing	0	0	0	0	0.30	0.30	0.40	$80,000
Purchasing	0.10	0.10	0.10	0	0.20	0.20	0.30	$10,000

fraction of the service departments' costs that is **assumed** to be assigned to maintenance, and so on.

The problem is to determine each service department's total costs (overhead plus charges for services provided by other departments) and to allocate these costs to the production departments.

Let x, y, z, and w denote the total costs in January of the accounting, maintenance, marketing, and purchasing departments, respectively. From column 1 we have that the costs of the maintenance, marketing, and purchasing departments that are assigned to accounting are $0.10y$, $0z$, and $0.10w$, respectively. The overhead of the accounting department (costs that are directly assigned to accounting such as salaries of employees in accounting, equipment, supplies, etc.) for January is $20,000. Thus x, the total cost of accounting, must satisfy the following condition:

$$x = 20{,}000 + 0.1y + 0.1w$$

A similar analysis for the maintenance, marketing, and purchasing departments yields the following conditions:

$$y = 18{,}000 + 0.02x \qquad\quad + 0.1w$$
$$z = 80{,}000 + 0.1\ x + \ 0.2y + 0.1w$$
$$w = 10{,}000 + 0.1\ x + \ 0.1y$$

Rearranging terms gives us the following system:

$$x - 0.1y \qquad - 0.1w = 20{,}000$$
$$-0.02x \ \ + y \qquad - 0.1w = 18{,}000$$
$$-0.1\ x - 0.2y + z \ - 0.1w = 80{,}000$$
$$-0.1\ x - 0.1y \qquad\ + w = 10{,}000$$

7.3   Traffic Network Flow

Part of a traffic network being designed to service a component of the Johnson City Airport is shown in Figure 7.1.

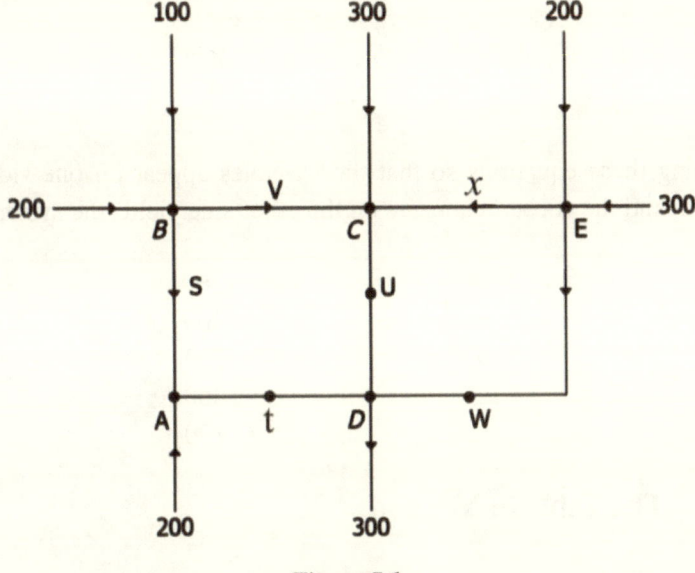

Figure 7.1

The roads are one way, as shown by the arrows, and the given values express the expected number of cars entering and leaving the network per hour during a heavy load period. The total number of cars entering the network, 800, equals the total number leaving the network, so that a basic traffic-flow equilibrium condition is satisfied. Another basic equilibrium condition is that the total number of cars entering each intersection point equal the total number of cars leaving the intersection point.

Assuming this to be the case, let us explore conclusions that can be drawn about traffic flow in the interior branches of the network.

To do so, we introduce variables s, t, u, v, w, and x as shown in Figure 7.1. Variable x denotes the number of cars passing between intersection points E and C per hour, t denotes the number of cars passing between intersection points A and D per hour, and so on. Since $200 + s$ cars enter A while t cars leave A, for equilibrium we have:

$$A: s + 200 = t$$

Similarly, for B, C, D, and E; we obtain:

$$B: \quad 200 + 100 = s + v$$
$$C: \quad v + x = u + 300$$
$$D: \quad t + u = w + 300$$
$$E: \quad w + 300 = x + 200$$

Rewriting these equations so that the variables appear on one side of an equation and the constant appears on the other side yields the system:

$$
\begin{array}{rcl}
s - t & & = -200 \\
s \quad + v & & = 300 \\
-u + v & + x & = 300 \\
t + u & - w & = 300 \\
& -w + x & = 100
\end{array}
$$

7.4   The Tableau Method

Since the elimination of a variable method, among others, that are usually employed to solve 2 by 2 systems and extended to 3 by 3 systems cannot efficiently be extended to larger systems, I employ what my colleagues Al Gewirtz, Lou Quintos and I termed the tableau method in an earlier book [1]. This method employs the same row and column operation used in matrix inversion, but without requiring the introduction of matrix algebra. Once students get a grip on the rhythm of the row and column operations employing the tableau method is relatively smooth sailing.

Its availability makes possible the discussion of serious applications of the kind illustrated by the afore examples. Discussion of the development of the tableau method can be found in [2; ch. 4] and [3; ch. 1].

For the systems of linear equations arising from the Arkin Company's problem and the Johnson City Airport traffic network problem the sequence of tableaus leading to their solutions are shown in Figures 7.2 and 7.3. Values to be converted to 1 are circled and to help you to keep track of the sequence of row operations carried out, the origin of each row is indicated by the notation on the right side of the tableaus. To handle the arithmetic arising from the numbers in the Austin Company's system of equations I used a primitive (by today's standards) hand calculator.

$$
\begin{aligned}
x - 0.1y \quad\quad - 0.1w &= 20{,}000 \\
-0.02\,x + \quad y \quad\; - 0.1w &= 18{,}000 \\
-0.1\; x - 0.2y + z - 0.1w &= 80{,}000 \\
-0.1\; x - 0.1y + \quad\quad\; w &= 10{,}000
\end{aligned}
$$

x	y	z	w	
①	-0.1	0	-0.1	20,000
-0.02	1	0	-0.1	18,000
-0.1	-0.2	1	-0.1	80,000
-0.1	-0.1	0	1	10,000

		x	y	z	w		
	1	①	-0.1	0	-0.1	20,000	
	2	-0.02	1	0	-0.1	18,000	
T_1	3	-0.1	-0.2	1	-0.1	80,000	
	4	-0.1	-0.1	0	1	10,000	row ①
	5	1	-0.1	0	-0.1	20,000	(0.02) row ⑤ + row ②
	6	0	(0.998)	0	-0.102	18,400	(0.1) row ⑤ + row ③
T_2	7	0	-0.21	1	-0.11	82,000	(0.1 row ⑤ + row ④
	8	0	-0.11	0	0.99	12,000	
	9	1	0	0	-0.1102204	21,843.69	(0.01) row ⑩ + row ⑤
	10	0	1	0	-0.1022044	18,436.87	$\left(\frac{1}{0.998}\right)$ row ⑥
T_3	11	0	0	1	-0.1314629	85,871.74	(0.21) row ⑩ + row ⑦
	12	0	0	0	(0.9787576)	14,028.06	(0.11). row ⑩ + row ⑧
	13	1	0	0	0	23,423.43	(0.11002204) row ⑯ + row ⑨
	14	0	1	0	0	19,901.72	(0.1022044) row ⑯ + row ⑩
T_4	15	0	0	1	0	87,755.93	(0.1314629) row ⑯ + row ⑪
	16	0	0	0	1	14,332.52	$\left(\frac{1}{0.9787576}\right)$ row ⑫

$$x = 23{,}423.43 \qquad\qquad z = 87{,}755.93$$

Figure 7.2

Accounting: $23,423
Maintenance: $19,902
Marketing: $87,756
Purchasing: $14,333

From these cost values and the percentages given in columns 5, 6, and 7 of Table 7.1, we obtain the following allocation of service departments' costs to the production departments:

Department P_1: $(0.24)(23,423) + (0.20)(19,902) + (0.30)(87,756)$
                     $+ (0.20)(14,333) = \$38,795$

Department P_2: $(0.26)(23,423) + (0.20)(19,902) + (0.30)(87,756)$
                     $+ (0.20)(14,333) = \$39,264$

Department P_3: $(0.28)(23,423) + (0.20)(19,902) + (0.40)(87,756)$
                     $+ (0.30)(14,333) = \$49,941$

Traffic Network Flow

$$
\begin{aligned}
s - t & & & & & = -200 \\
s + & & v & & & = 300 \\
& -u + v & & +x & & = 300 \\
t + u - & & w & & & = 300 \\
& & & -w + x & & = 100
\end{aligned}
$$

s	t	u	v	w	x	
①	-1	0	0	0	0	-200
1	0	0	1	0	0	300
0	0	-1	1	0	1	300
0	1	1	0	-1	0	300
0	0	0	0	-1	1	100

		s	t	u	v	w	x		
	①	①	−1	0	0	0	0	−200	
	②	1	0	0	1	0	0	300	
T_1	③	0	0	−1	1	0	1	300	
	④	0	1	1	0	−1	0	300	
	⑤	0	0	0	0	−1	1	100	
	⑥	1	−1	0	0	0	0	−200	row ①
	⑦	0	①	0	1	0	0	500	(−1)row ⑥ + row ②
T_2	⑧	0	0	−1	1	0	1	300	row ③
	⑨	0	1	1	0	−1	0	300	row ④
	⑩	0	0	0	0	−1	1	100	row ⑤
	⑪	1	0	0	1	0	0	300	row ⑫ + row ⑥
	⑫	0	1	0	1	0	0	500	row ⑦
T_3	⑬	0	0	−1	1	0	1	300	row ⑧
	⑭	0	0	①	−1	−1	0	−200	(−1)row ⑫ + row ⑨
	⑮	0	0	0	0	−1	1	100	row ⑩
	⑯	1	0	0	1	0	0	300	row ⑪
	⑰	0	1	0	1	0	0	500	row ⑫
T_4	⑱	0	0	0	0	−1	1	100	row ⑲ + row ⑬
	⑲	0	0	1	−1	−1	0	−200	row ⑭
	⑳	0	0	0	0	⊖	1	100	row ⑮
	㉑	1	0	0	1	0	0	300	row ⑯
	㉒	0	1	0	1	0	0	500	row ⑰
T_5	㉓	0	0	0	0	0	0	0	row ㉕ + row ⑱
	㉔	0	0	1	−1	0	−1	−300	row ㉕ + row ⑲
	㉕	0	0	0	0	1	−1	−100	(−1)row ⑳

Figure 7.3

Writing the equations that correspond to tableau T_5 yields:

$$s \quad\;\; +v \qquad\quad\; = 300$$
$$t \;\; +v \qquad\quad\; = 500$$
$$u - v \quad\; - x = -300$$
$$w - x = -100$$

By transposing the terms arising from columns in T_s that are not in zero-one form (v and x columns), we obtain the following description of the solutions of our system.

$$s = 300 - v$$
$$t = 500 - v$$
$$u = -300 + v + x$$
$$w = -100 + x$$
$$v \text{ is arbitrary}$$
$$x \text{ is arbitrary}$$

Although this system has infinitely many solutions, only a small number of them make sense in terms of the network. Since the variables express the number of cars passing per hour between branches connecting intersection points of the network, the values given to these variables must, at the very least, be restricted to nonnegative integers (0, 1, 2, etc.). The requirement $s \geq 0$ yields:

$$s = 300 - v \geq 0, v \leq 300$$

From $w \geq 0$ we have:

$$w = -100 + x \geq 0, x \geq 100$$

From $u \geq 0$ we obtain:

$$u = -300 + v + x \geq 0, v + x \geq 300$$

Thus traffic equilibrium cannot be maintained if more than 300 cars per hour pass between B and C, or fewer than 100 cars per hour pass between E and C, or the sum of the number of cars passing per hour between B and C and E and C is less than 300.

This tells us something about the practical feasibility of the network.

7.5   Expanding Student Horizons

I believe that introducing students to the two scenarios considered (to take two examples) and the tableau method serve as useful tools for expanding student horizons in a number of important ways.

1. The scenarios can be employed to serve as a vehicle, with realistic overtones, for discussion of math modeling or restricted to systems of linear equations, depending on circumstances.

2. The systems of linear equations that emerge illustrate at least two noteworthy properties:

    (a) Real world problems may give rise to larger than usual systems that are the focus of "standard" algebra courses (2 by 2 and 3 by 3 systems) with, how should I put it, messy, challenging coefficients (as opposed to 2, 3, -2, 5 *et al*).

    (b) Systems of linear equations with infinitely many solutions exist and may have to be dealt with in terms of describing them and giving thought to which ones make sense in terms of the application under study.

3. From these examples the versatility of the tableau method for solving systems of linear equations is made clear. It is applicable in the same way no matter the size of the system (4 by 4, 40 by 40, 20 by 60 *et al*) and the nature of the solutions (none, one, infinitely many).

4. Since the row operations that underlie the tableau method are based on a straightforward routine, the computer can be brought into play to execute it. If the computer dimension is available to students, this provides a suitable setting to illustrate its application.

5. Since the row operations that underlie the tableau method also underlie matrix inversion and the simplex method for solving linear programs, the tableau method can serve as a door opener to these topics by providing a straightforward setting where it is clear what the row operations do in terms of linear equations.

7.6   Food for Thought Questions

1. The Sonin Company, which makes computers, has two production departments, P_1 and P_2, and three service departments, S_1, S_2, and S_3. Each service department's total cost must be distributed to the production

departments and to the other service departments based on their respective usages of the services provided. For each service department listed in the leftmost column of Table 7.2, the **assumed** fraction of its total cost assigned to the service and production departments of the firm is given.

Table 7.2

SERVICE DEPARTMENT	SERVICE DEPARTMENT			PRODUCTION DEPARTMENT		MARCH OVERHEAD
	S_1	S_2	S_3	P_1	P_2	
S_1	0	0.10	0.05	0.40	0.45	$40,000
S_2	0.10	0	0.10	0.40	0.40	$30,000
S_3	0.20	0.05	0	0.35	0.40	$20,000

Set up the system of equations that describes the conditions to be satisfied by the total costs of the service departments.

2. Part of a traffic network being designed to service an envisioned shopping center is shown in Figure 7.4.

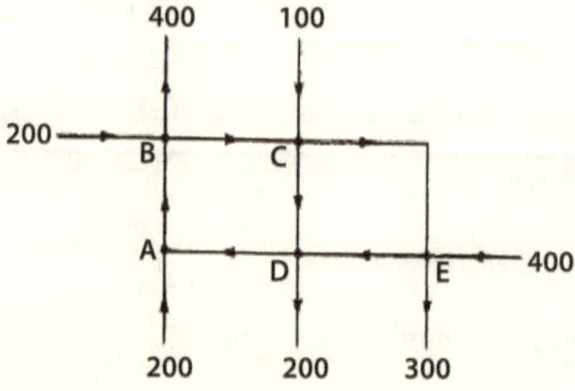

Figure 7.4

The total number of cars entering the network per hour during a peak period, 900, equals the total number leaving the network per hour, so that a basic equilibrium condition is satisfied.

On the basis of the equilibrium condition that the total number of cars entering each intersection point equals the total number leaving the intersection point, set up a system of linear equations to describe traffic flow in the interior of the network.

7.7   References of Interest?

1. W. J. Adams, A. Gewirtz, L. V. Quintas, *Elements of Linear Programming* (New York: Van Nostrand Reinhold Co., 1969)

2. William J. Adams, *Algebra with Applications: Technique and THOUGHT*, Revised Edition (Philadelphia: Xlibris, 2009). Available on the web at webpage.pace.edu/wadams.

3. William J. Adams, *Finite Mathematics, Models, and Structure*, Revised Edition (Philadelphia: Xlibris, 2009). Available on the web at webpage. pace.edu/wadams.

8

Math Modeling to Achieve Profit Maximization

8.1   Preface

The Gaja Company, considered in ch. 4, sec.s 4.2 and 4.6 produces shoes. Two new models, tentatively designated A-18 and A-21, are to be introduced into the market. Based on two math production scheduling models, M1 and M2, the projected monthly profit is $230,000 and $200,000, respectively. Should M1 be implemented because its projected profit is larger than M2's? The answer, of course, is NO. This decision making criterion is far, far too simplistic, and choosing a production scheduling model on this basis might have disastrous consequences for the Gaja Company.

A return to the Austin Company scenario presented in the Introduction provides us with a platform for formulating realistic decision making criteria for choosing a production scheduling model for implementation.

8.2   The Austin Company's Profit Maximization Problem

From the Introduction:

The Austin Company, a producer of high quality electronic home entertainment equipment, has decided to enter the digital tape player market by introducing two models, Ultra (DT-1) and Supreme (DT-2). Their problem is to determine the number of units of each model that should be produced weekly to maximize profit.

The Company's operations research department was asked to study the problem and make recommendations. The OR department began its analysis by collecting data. They viewed the manufacturing process in terms of three phases; construction, assembly, and finishing. The data collected and their analysis led them to introduce the following **assumptions**, which they took as **postulates**.

P1:   In the construction phase each DT-1 unit requires 2 hours of labor and each DT-2 unit requires 3 hours of labor. At most 1,100 hours of construction time are available per week.

P2:   In the assembly phase each DT-1 unit requires 5 hours of labor and each DT-2 unit requires 3 hours of labor. At most 1,400 hours of assembly time are available per week.

P3:   In the finishing phase each DT-1 unit requires 4 hours of labor and each DT-2 unit requires 1 hour of labor. At most 756 hours of finishing time are available per week.

P4:   After taking the cost and revenue factors into consideration the anticipated profit for each DT-1 unit is $150 and the anticipated profit for each DT-2 unit is $120. In order for these unit profit values to be realistic the Company must produce at least 25 DT-1 and 40 DT-2 units per week.

P5:   There is an unlimited market for the DT-1 and DT-2 models.

P: Other factors that may play a role in determining the production schedule that would maximize profit are not being considered. (This may be for various reasons. They may view some factors as negligible to profit maximization. There may be factors whose importance they recognize, but do not know how to handle. And then, possibly, there are factors that are important, but whose significance they have not recognized.)

Their next task was to translate these **postulates** into mathematical form, being careful to include everything stated in them but not go beyond what is being assumed. The OR department began by introducing variables for the quantities it sought to determine; they let x denote the number of DT-1 and y the number of DT-2 units to be made weekly.

The afore conditions led them to the math problem:

$$\text{Maximize } P(x, y) = 150x + 120y$$

subject to

$$x \geq 0, y \geq 0$$
$$2x + 3y \leq 1100$$
$$5x + 3y \leq 1400$$
$$4x + y \leq 756$$
$$x \geq 25, y \geq 40$$

For ease of reference we name this mathematical structure together with the **postulates** that underlie it linear program model LP-1.

The Austin Company also hired the Aleksa Company, a consulting operations research firm, to independently study their digital tape player problem and make recommendations. The Aleksa OR group viewed the manufacturing process in terms of two phases: construction (which included assembly) and finishing. The data collected and their analysis led them to introduce the following **assumptions/postulates**.

P1a: In the construction phase each DT-1 unit requires 8 hours of labor and each DT-2 unit requires 5 hours of labor. At most 2,210 hours of construction time are available per week.

P2a: In the finishing phase each DT-1 unit requires 3 hours of labor and each DT-2 unit requires 2 hours of labor. At most 860 hours of finishing time are available per week.

P3a: The anticipated profit for each DT-1 unit is $140 and the anticipated profit for each DT-2 unit is $150. In order for these unit profit values to be realistic the company must produce at least 50 DT-1 and 50 DT-2 units per week.

P4a: There is an unlimited market for the DT-1 and DT-2 models.

P: Other factors that may play a role in determining the production schedule that would maximize profit are not being considered.

The afore conditions led the Aleksa Company to the math problem:

$$\text{Maximize } P(x, y) = 140x + 150y$$

subject to
$$x \geq 0, y \geq 0$$
$$8x + 5y \leq 2210$$
$$3x + 2y \leq 860$$
$$x \geq 50, y \geq 50$$

where x represents the number of DT-1 and y the number of DT-2 units to be made weekly.

For ease of reference we name this mathematical structure together with the **postulates** that underlie it LP-2.

For discussion of details on the formulation of LP-1 or LP-2 see [1; ch. 6] or [2; ch. 3].

8.3   The Linear Program and Linear Program Model

A **linear program** is a mathematical problem with the following structure: there is specified a linear function of a number of variables that are required to satisfy linear conditions described by some mixture of linear inequalities and linear equations, called **constraints**. The problem is to find values for these variables which satisfy the constraints and yield the maximum, or minimum, value of the function, which is called an **objective function**.

A linear program may or may not arise from a real world situation/problem under study. If it does, the linear program together with the **assumptions/ postulates** that led to it is called a **linear program model** for the situation/ problem under study.

LP-1 and LP-2 are 2-variable linear program models, but the same kind of problem may involve 200 variables, 2000 variables or even 200,000 or more variables.

8.4   The Corner Point Solution Method

In the Introduction it is noted that the solutions of LP-1 and LP-2 are (100, 300) with maximum value 51,000 and (50, 355) with maximum value 60,250, respectively. For 2-variable linear program models the corner point method is a (relatively) simple solution technique for such problems. Considering the focus of this book it is not necessary to consider its details here.

As previously, for those who are interested in pursuing it I call your attention to [1; ch. 6] and [2; ch. 3].

8.5   The Austin Company's Question: Which Solution, if Either, Should Be Implemented?

This, needless-to-say, is the key issue in every math application situation. You have the solution, or solutions to make it more interesting and insightful, and the crown jewel of all this effort is the question, now what; **which solution should be implemented?** (Alas, this crown jewel question is usually completely ignored in algebra courses.)

> It's important, I believe, to add the conditional, **if either**, to the question. It's so tempting to jump to the erroneous conclusion that one must accept one or the other of the solutions obtained for implementation.

I employ the following kind of scenario to help bring out these points:

It's always easier when you have a choice of one; take it, or leave it. But a choice of two is another matter. Bottom-line Bob, chairman of the ten member board charged with making a decision on how to implement the Company's entry into the digital tape player market, argued that it's obvious what we should do. "Implementation of LP-2 brings us a weekly profit of $60,250, whereas implementation of LP-1 brings us a weekly profit of $51,000. Since we want the largest possible return, we should go with LP-2. It's a no-brainer. The board voted nine to one to implement LP-2.

Alas, the $60,250 weekly profit was far from being realized after LP-2 was implemented, and two years later the Austin Company's venture into the digital tape player market had to be written off as a disaster.

Bottom-line Bob, who presided over this disaster was confused, upset, and out of a job. He went to Reflective Ramunė, the chair of the new board and the one person who had voted against implementation of LP-2, with some questions: "We were ultra-cautious and obtained the additional services of the Aleksa OR group to make recommendations which, subsequently, had disastrous consequences for us; what went wrong? How is it that mathematics failed us? Why did you vote against implementation of LP-2?"

"Well Bob, as I pointed out at the board meeting, I voted against implementation of LP-2 because I was not convinced that its promise of a $60,250 weekly profit was realistic. As you yourself pointed out, 'the promise of LP-2 is $9,250 better than that promised by LP-1,' but promises may not be realizable if they are founded on unrealistic assumptions. The conclusion reached from LP-2 was indeed tempting, and in fact proved too tempting for my colleagues on the board, but since it came from a linear program founded on **assumptions** which I viewed as unrealistic, I resisted temptation. We have no quarrel with mathematics; mathematics gave us a valid conclusion from LP-2, which is all that we can legitimately expect. Unfortunately that conclusion proved to be unrealistic."

Is Mathematics Precise?

"Ramunė, I don't understand this. I always liked math in high school and college. Solving those equations, factoring those expressions, differentiating those functions, throwing the data into the computer and letting it do its thing, that was real fun. What I like most about math is its *precision*. You don't get ten sides to a story. You get one answer and that's that; no baloney.

"Bob, I think your math courses may have focused too much on technique and not enough on perspective. Technique can be fun to a point, but without a perspective

on its place in the over-all role of mathematics in applications, we see only a small tip of the mathematical iceberg. Mathematics is precise in the sense that it gives us valid conclusions based on the **assumptions** made, which is where technique—factoring, differentiating functions, and the like—plays its major role.

> Whether the **assumptions** made are realistic or not is another matter which technique can't help us with. The question of how to formulate these **assumptions** and reach a judgment on their realism may indeed yield ten sides to the story. I'm afraid that those who find mathematics attractive because of what they perceive to be its absolutist nature have misunderstood the meaning of mathematical *precision*."

8.6   How Could It Be Wrong? I Used a Computer

"Ramunė, I still don't understand what went wrong. The company just spent millions to update its computer system. I had access to the latest and the best. Why didn't this save us from disaster." "Bob, Henry Clay's observation that 'statistics are no substitute for judgment' applies equally well to the computer.

We cannot expect the computer to employ technological alchemy and convert unrealistic assumptions into golden truths. Keep in mind the NINO principle; if nonsense in, then nonsense out. Indiscriminate use of computer technology has made possible the generation of more and more nonsense more quickly than ever before by more people having less and less understanding of what they are doing." (Rasa, let us recall, was burned by NINO. (ch. 2, sec.s 2.6, 2.8.))

The Computer's Right of Way

"If what you say is true Ramunė, then what good is this super computer technology to us?" "For number crunching and delivering results quickly and efficiently, the computer is without equal, Bob.

In this dimension it is the undisputed master of the field. The mathematical model building process and computers have developed a symbiotic relationship in that computers have made it possible for us to solve previously unapproachable large scale problems that come out of mathematical models, while the accessibility of such

problems to computer solution has made possible the use of such complex models. Alas, none of this overrides the NINO principle."

8.7   Unity Born of Diversity in the Linear Programming Landscape

To show off the unity that the linear program model brings to a variety of problems that differ considerably in their outer appearance (illustrated by Figure 8.1) I take up with my students the following illustrations, among others,

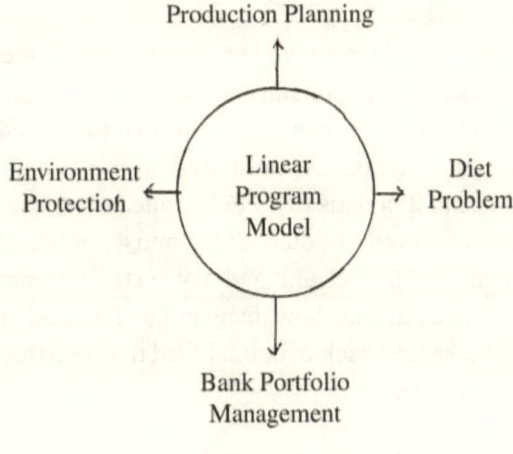

Figure 8.1

Case 1 Production Planning

The Austin Company's problem of determining the number of DT-1 and DT-2 digital tape players to be made per week so as to maximize profit, considered in sec. 8.2, is a production scheduling problem which we expressed in linear program terms under the **assumptions** introduced. The background that led to the Austin Company's linear program models is illustrative of situations with the following general features:

A firm makes a number of products or models of a product and utilizes a number of resources in their manufacture, such as raw materials, labor, capital, different machines, storage facilities. It is assumed that for each product made a fixed amount of each resource is required to make a unit of that product. Within the production time frame a fixed amount of each resource is available and cannot be exceeded. It is also **assumed** that for a range of possible output levels there is a fixed profit per unit of each product which does not depend on the number of units produced.

Under these conditions the problem of determining output levels of the products produced so as to maximize total profit can be formulated in terms of a linear program model.

Case 2 Diet Problems

A sack of animal feed is to be put together from linseed oil meal and hay. It is required that each sack of feed contain at least 2 pounds of protein, 3 pounds of fat, and 8 pounds of carbohydrate. It is **estimated (assumed,** if you prefer) that each unit (a unit is 30 pounds) of linseed oil meal contains 1 pound of protein, 1 pound of fat, 2 pounds of carbohydrate, and that each unit of hay contains 1/2 pound of protein, 1 pound of fat, and 4 pounds of carbohydrate. Linseed oil meal costs $1.50 per unit and hay costs $1.10 per unit.

The problem is to determine how many units of linseed oil meal and hay should be used to make up a sack of animal feed that satisfies the nutritional requirements at minimal cost.

This problem illustrates "diet problems" with the following general features: A diet, or food substance, is to be put together from a number of available foods. It is required that the diet be balanced in the sense that it must contain minimal amounts of stated nutrients—proteins, fats, carbohydrates, minerals, vitamins, etc. It is **assumed** that each food unit contains a known fixed amount of each nutritional unit and that the unit prices of the food items are known and fixed within the time period considered. The problem is to determine the minimal cost diet which satisfies the prescribed nutritional requirements.

Case 3 Environmental Protection

The Saxon Company must produce at least 250 thousand tons of paper annually. From the current operating system 10 pounds of chemical residue is deposited into a neighboring water system for each ton of paper produced. The resulting pollution has become a problem of serious concern, and to remain eligible for state tax benefits the Saxon Company must restrict the chemical residue emitted into the state's water system to not exceed 200 tons per year. Two filtration systems, Delta and Beta, have emerged for consideration. It is **estimated (assumed)** that the installation of the Delta system would reduce emissions to 2 pounds for each ton of paper produced, and installation of the Beta system would reduce emissions to 1 pound for each ton of paper produced. Capital and operating costs for the Delta and Beta systems have been **estimated (assumed to be)** at $8 and $12, respectively, per ton of paper produced.

The problem is to determine how many tons of paper should be produced subject to the Delta system and how many should be produced subject to the Beta system so that the emissions standard is met at minimal cost.

Case 4 Bank Portfolio Management

The Charles National Bank has assets in the form of loans and negotiable securities which, it is **assumed**, bring returns of 10 and 8 percent, respectively, in a certain time period. The bank has a total of $60 million to allocate between loans and securities. To meet unanticipated deposit withdrawals the bank maintains a securities balance greater than or equal to 25 percent of total assets. Lending is the bank's most important activity and to satisfy its clients it requires that at least $15 million be available for loans.

The bank wishes to determine, under these conditions, how funds should be allocated to maximize total investment income. (see [3].)

The afore case studies are realistic, I point out to my students, but are presented in miniature for the sake of classroom manageability.

Actual real-life situations that emerge have the same structure and tone, but are more complex in that more factors are generally considered and more variables are required.

We view all of such situations through the eyes of others in much the same way that we see events through the eyes of a reporter or observer by reading his account of them in a newspaper, journal or book. Just as the reporter has selected what he believes are important features surrounding the events and has omitted those he considers unessential, we too are looking at features considered crucial to the situations we examine as seen by someone who has made such a selection. This selection reflects **assumptions** that have been made. To maintain a proper perspective on this it is important to keep in mind that other analysts, as other reporters, might see things in a different light and accordingly make other **assumptions/postulates**.

8.8   Food for Thought Questions

1. The Veronika Company makes stereo systems. Two new models, RA5 and RA9, are to be mass produced. Both models pass through assembly and finishing plants of the company. In the assembly plant an RA5 unit is worked on for 1 hour; an RA9 unit is worked on for 3 hours. In the finishing plant an RA5 unit is worked on for 2 hours; an RA9 unit is worked on for 1 hour. At most, 90 hours of assembly time and 80 hours of finishing time are available per week. The **anticipated (assumed)** profit on an RA5 unit is $10 and the **anticipated (assumed)** profit on an RA9 unit is $15.

    The problem is to determine, with respect to the given **assumptions/ postulates**, how many RA5 and RA9 units should be made per week so as to maximize profit.

    We begin by introducing variables to stand for the quantities we wish to determine. Let x denote the number of RA5 units to be made and let y denote the number of RA9 units to be made. To make needed information available at a glance, we express the basic data in tabular form, as shown in Table 8.1

Table 8.1

	No. of units to be made per week	Profit per unit	Assembly time per unit (hours)	Finishing time per unit (hours)
Model RA5	x	$10	1	2
Model RA9	y	$15	3	1

Since profit is to be maximized, we must express profit in terms of x and y. It is useful to note that:

$$profit = \begin{bmatrix} profit\ on \\ model\ RA5 \end{bmatrix} + \begin{bmatrix} profit\ on \\ model\ RA9 \end{bmatrix}$$

$$profit = \begin{bmatrix} profit\ on \\ one\ RA5 \\ unit \end{bmatrix} \cdot \begin{bmatrix} no.\ of \\ units \\ made \end{bmatrix} + \begin{bmatrix} profit\ on \\ one\ RA9 \\ unit \end{bmatrix} \cdot \begin{bmatrix} no.\ of \\ units\ made \end{bmatrix}$$

$$profit = 10x + 15y$$

The profit obtained by making x RA5 units and y RA9 units is expressed by the linear function

$$P(x, y) = 10x + 15y,$$

where x and y denote the number of RA-5 and RA-9 units to be made per week, respectively.

Our next task is to describe the conditions that x and y must satisfy. Since the number of units made is nonnegative, we have:

$$x \geq 0$$
$$y \geq 0$$

To express assembly plant operation time in terms of x and y, we note that:

$$\begin{bmatrix} assembly \\ plant \\ time \end{bmatrix} = \begin{bmatrix} assembly \\ time\ on \\ RA5 \end{bmatrix} = \begin{bmatrix} assembly \\ time\ on \\ RA9 \end{bmatrix}$$

$$\begin{bmatrix} assembly \\ plant \\ time \end{bmatrix} = \begin{bmatrix} assembly \\ time\ for \\ 1RA5\ unit \end{bmatrix} \cdot \begin{bmatrix} no.of \\ RA5\ unit \\ made \end{bmatrix} + \begin{bmatrix} assembly \\ time\ for \\ 1RA9\ unit \end{bmatrix} \cdot \begin{bmatrix} no.of \\ RA9\ units \\ made \end{bmatrix}$$

$$\begin{bmatrix} assembly \\ plant \\ time \end{bmatrix} = 1 \cdot x + 3 \cdot y$$

Since at most 90 hours of assembly time per week is available, we have

$$x + 3y \le 90.$$

Similarly, the condition that at most 80 hours of finishing time are available per week is expressed in terms of x and y by the inequality

$$2 + y \le 80.$$

In summary, the **postulates** introduced lead to the linear program:

$$\text{Maximize } P(x, y) = 10x + 15y$$

subject to
$$x \ge 0$$
$$y \ge 0$$
$$x + 3y \le 90$$
$$2x + y \le 80$$

Application of the corner point method yields solution (30, 20) and maximum value 600. Should this solution be implemented? William Ganz, production manager of the Veronika Company gave thought to this question and asked for opinions.

1. (a)   "Yes," said Jim Turner, the resident computer "expert". "The solution was obtained by use of the latest computer technology available, and that's good enough for me." Would you agree with Jim? Explain (Red Herring Alert.)

    (b)   "Of course, no question about it," said June Carver, CEO of the Veronika Company. "The corner point method is the mathematical basis for this solution and its use ensures that profit will be maximized." Would you agree with June? Explain. (Red Herring Alert.)

    (c)   "Are there any factors that we should have taken into consideration in the formulation of this linear program model," Ganz asked Horace Black, head of the operations research department. "None that we could come up with," replied Horace. "We took into account every factor that we judged relevant to maximization of profit." "What about the linear program models formulated for the Austin Company? Are there lessons that we should give thought to from these models?" "I don't think so," replied Horace. Would you agree with Horace's assessment? Explain.

    If you were asked for your thoughts on whether the solution (30, 20) should be implemented, what would you say to Mr. Ganz? Explain.

2. The Onutė Corporation plans to introduce two high resolution TV models, T20 and T24, to the market. Its own operations research group was led to introduce the following M1 model to determine the optimal production schedule for maximizing profit:

    $$\text{Maximize } P(x, y) = 180x + 120y$$
    subject to
    $$x \geq 0, y \geq 0$$
    $$4x + 3y \leq 320$$
    $$5x + 2y \leq 330,$$

    where, x and y denote the number of T20 and T24 units, respectively, to be made daily. Its solution is (50, 40) with maximum value 13,800.

The Aleksa company was also hired to study the Onutė Corporation's production scheduling problem. It was led to introduce the following M2 model to determine the optimal production schedule for maximizing profit:

$$\text{Maximize } P(x, y) = 190x + 110y$$

subject to

$$x \geq 0, y \geq 0$$
$$5x + 2y \leq 330$$
$$3.25x + 2y \leq 225$$
$$4x + 3y \leq 320,$$

where x and y denote the number of T20 and T24 units, respectively, to be made daily. Its solution is (60, 15) with maximum value 13,050.

The following questions have arisen. How would you answer them?

(a)   If mathematics is the *precise* subject that it is reputed to be, should there not be one solution to this problem rather than two?

(b)   Since two solutions emerge, does it follow that not both are valid? Explain.

(c)   Before making a decision about whether to implement M1 or M2, what questions would you put to the two operations research groups?

(d)   Which model, if either, would you adopt and implement? Why? Is it possible that you would not adopt either model?

3.   The Atlantic Company, a lamp manufacturer, began production of two models, A14 and A51. To determine the best production schedule, two consulting firms, the Aleksa Company and Veronika Consultants, were hired to analyze the company's operations and make recommendations. The Aleksa Company set up a linear program model for the production process, which when solved by a mathematical method called the simplex method yielded the solution (500, 280) with a maximum value of 3000. The Aleksa Company recommended that 500 A14 units and 280 A51

units be made per week, for an anticipated maximum profit of $3000 per week.

Veronika Consultants set up a different linear program model for the production process, which when solved by the corner-point method yielded (450, 300) as the solution with a maximum value of 2500. Veronika Consultants recommended that 450 A14 units and 300 A51 units be made weekly to obtain a maximum profit of $2500 per week.

The management of the Atlantic Company found these developments puzzling and raised the following questions. Answer these questions in appropriate detail.

(a)   How is it possible for different solutions to be obtained? After all, isn't mathematics a *precise* subject?

(b)   Which solution is *correct* and in what sense is it *correct*?

(c)   Which solution should be implemented and why? Is it possible that neither solution should be implemented? Explain.

8.9   Ideology vs. Mathematics: From Ideologically Incorrect to Nobel Prize

Numbers 5 and 6 of math insight, myth, in a sense a bit of both or nonsense, stated in the Introduction notes:

• The *precision* of mathematical methods guarantees unassailable conclusions which serve as pillars of stability and strength in a world besieged by foggy thinking, prejudice, and rampant special interests.

• Mathematically derived conclusions are indisputable because they are based on deductive logic, which is untainted by bias and ideology.

Your first thought might be to judge them math insights. In the math-world itself, perhaps yes, but the real-world is another matter. The early history of linear programming gives us an interesting case. Another interesting case is found in math model building for the Consumer Price Index in the 1990s; ch. 10, sec. 10.8.

The seed from which linear programming first germinated was planted in the late 1930s when the Leningrad Plywood Trust approached the Mathematics and Mechanics Department of Leningrad University for help in solving a production scheduling problem of the following nature. The Plywood Trust had different machines for peeling logs for the manufacture of plywood. Various kinds of logs were handled and the productivity of each kind of machine (that is, the number of logs peeled per day) depended on the wood being worked on. The problem was to determine how much work time each kind of machine should be assigned to each kind of log so that the number of peeled logs produced is largest. A basic condition which had to be satisfied is that if logs of a given type of wood, oak, let us say, made up a specified percent of the input, 5 percent, for example, the peeled oak logs would also make up 5 percent of the output.

The germination of this seed is due to Leonid Kantorovich, who saw that it together with a wide variety of economic planning problems can be formulated in terms of math problems called linear programs. These problems involved the optimum distribution of worktime of machines, minimization of scrap in manufacturing processes, best utilization of raw materials, optimum distribution of arable land, optimal fulfillment of a construction plan with given construction materials, and the minimal cost plan for shipping freight from given sources to given destinations.

In 1939 Kantorovich published a report [4] on his discoveries which included a method for solving all the linear programs he had formulated for the aforenoted problems. The chaos of the Second World War and the postwar intellectual climate in the Soviet Union did not allow for the development and implementation of Kantorovich's linear programming methods in the Soviet economic scene. Any works that seemed to deviate from Marxist-Leninist ideology as interpreted by Joseph Stalin's closest ideologues could have chilling consequences for its authors. Kantorovich's mathematical methods for economic planning problems were neutral as far as Marxist-Leninist thinking was concerned, but they were new and a radical departure from comfortable orthodoxy, which made them suspect. This was enough to put them into a deep freeze.

The thaw in the Soviet Union's intellectual climate which followed Stalin's death in 1953 saw the rebirth, development, and implementation of Leonid Kantorovich's linear programming methods into the economic life of the U.S.S.R. Independently of the Soviet scene, linear programming methods were developed in the United States and Western Europe in the late 1940's, and the 1950's and 60's saw the development of a wide variety of linear program models for problems arising in such areas as economic planning, accounting, banking, finance, industrial engineering, and marketing.

In 1975 Kantorovich was a co-recipient of the Nobel Proze in economics for his development of linear programming methods and their application to economic planning.

8.10   References of Interest?

1. William J. Adams, *Algebra with Applications: Technique and THOUGHT*, Revised Edition (Philadelphia: Xlibris, 2009), ch. 6. Available on the web at webpage.pace.edu/wadams.

2. William J. Adams, *Finite Mathematics, Models, and Structure*, Revised Edition (Philadelphia: Xlibris, 2009), ch. 3. Available on the web at webpage.pace.edu/wadams.

3. A. Broaddus, "Linear programming: A New Approach to Bank Portfolio Management," *Federal Reserve Bank of Richmond: Monthly Review,* vol. 58, No. 11 (Nov. 1972), pp. 3-11. This article provides an introductory nontechnical discussion of linear programming for bank portfolio management.

    K. J. Cohen and F. S. Hammer, "Linear Programming and Optimal Bank Asset Management Decisions," *Journal of Finance,* vol. 22 (May 1967), pp. 147-165. This paper describes a linear program model that had been used for several years by Bankers Trust Company in New York to assist in reaching portfolio decisions.

4. L. V. Kantorovich, *Mathematical Methods of Organizing and Planning Production,* Leningrad University, 1939. For an English translation, see *Management Science,* vol. 6, no. 4 (July 1960), pp. 363-422; or V. S. Nemchinov, ed., *The Use of Mathematics in Economics* (Cambridge, Mass.: MIT Press, 1964).

9

Math Modeling for Judgments
Expressed in Quantitative Form

9.1   Preface

A number of problems lead to linear program models called 0-1 integer program models because the variables are restricted to the values 0 and 1. Two problem types, the Assignment Problem and Knapsack Problem, are considered.

9.2   The Assignment Problem

The Brian Publishing Company has two positions to fill, editor of the mathematics list (job 1) and editor of the social science list (job 2), and is considering three candidates, Albert Roberts (candidate 1), Rita O'Brien (candidate 2), and Martin Thorp (candidate 3). After considering resumes, letters of recommendation, and conducting interviews the editorial board of the company assigned a numerical rating to each person's qualifications for each position as stated in Table 9.1. These ratings

Table 9.1

	Candidate	Position	
		Math, Editor (job 1)	*Soc. Sci. Editor (job 2)*
Roberts	(candidate 1)	8	8
O'Brien	(candidate 2)	7	9
Thorp	(candidate 3)	9	8

serve as a quantitative measure of each candidate's potential for each position as seen by the editorial board. The editorial board wishes to assign candidates to positions in such a way that total potential is maximized.

To relate the candidates to the jobs we introduce X_{ij} to relate candidate i to job j. X_{ij} can assume one of two values, 0 if candidate i is not assigned job j, and 1 if candidate i is assigned job j. In summary, we emerge with Table 9.2.

Table 9.2

	Candidate	Position	
		Math, Editor (job 1)	*Soc. Sci. Editor (job 2)*
Roberts	(candidate 1)	X_{11}	X_{12}
O'Brien	(candidate 2)	X_{21}	X_{22}
Thorp	(candidate 3)	X_{31}	X_{32}

The function

$$P = 8X_{11} + 8X_{12} + 7X_{21} + 9X_{22} + 9X_{31} + 8X_{32},$$

obtained by multiplying the variable that relates a candidate to a job by the candidate's potential for the job and adding, is the **potential function to be maximized** subject to two conditions:

---

1. Each candidate is assigned to at most one job.

The variables X_{11} and X_{12} (row 1 of Table 9.2) relate candidate 1 to the available jobs 1 and 2. The constraint

$$X_{11} + X_{12} \leq 1$$

expresses the requirement that candidate 1 be assigned to at most one job since it makes impossible for candidate 1 to be assigned job 1 ($X_{11} = 1$) and job 2 ($X_{12} = 1$). This constraint comes from row 1 of Table 9.2, and in general the condition that each candidate be assigned at most one job is expressed by requiring that the sum of the variables in each row of Table 9.2 be less than or equal to one. Rows 2 and 3 yield the same condition for candidates 2 and 3:

$$X_{21} + X_{22} \leq 1$$
$$X_{31} + X_{32} \leq 1$$

---

2. Each job is filled by at most one person.

The variables X_{11}, X_{21}, and X_{31} in column 1 of Table 9.2 relate candidates 1, 2, and 3 to job 1. The constraint

$$X_{11} + X_{21} + X_{31} \leq 1,$$

obtained by requiring that the sum of the variables in the first column 1 of Table 9.2 be less than or equal to one, expresses the requirement that job 1 be filled by at most one candidate since it makes impossible for any two or all three candidates to be assigned to job 1 ($X_{11} = 1, X_{21} = 1, X_{31} = 1$). Column 2 yields the same condition for job 2:

$$X_{12} + X_{22} + X_{32} \leq 1$$

We thus obtain the following integer program model: Find nonnegative integers (zeros and ones) that

$$\text{Maximize } P = 8X_{11} + 8X_{12} + 7X_{21} + 9X_{22} + 9X_{31} + 8X_{32}$$

subject to

$$
\begin{aligned}
X_{11} + X_{12} &\leq 1 \\
X_{21} + X_{22} &\leq 1 \\
X_{31} + X_{32} &\leq 1 \\
X_{11} + X_{21} + X_{31} &\leq 1 \\
X_{12} + X_{22} + X_{32} &\leq 1,
\end{aligned}
$$

where $X_{11}, X_{12}, \ldots$ are defined in Table 9.2.

By inspection we can see from Table 9.1 that the potential function P is maximized when candidate 3 (Thorp) is assigned to job 1 (Math Editor) and candidate 2 (O'Brien) is assigned to job 2 (Social Science Editor). That is, $X_{11} = 0, X_{12} = 0, X_{21} = 0, X_{22} = 1, X_{31} = 1, X_{32} = 0$ maximizes the potential function. When the number of candidates and jobs is large such problems cannot be handled by inspection, but can be handled by integer programming methods.

In discussing this scenario in class I make it a point to reaffirm that the afore solution is a valid conclusion **with respect to** the quantitative measures of the candidates' potential ratings arrived at by the editorial board's **postulates,** no more and no less.

Assignment problems may involve the most efficient assignment of people to jobs, machines to tasks, project leaders to projects, police cars to city sectors, departments to store locations, sales people to territories, and so on. The objective might involve maximizing effectiveness in some sense or minimizing cost or travel time.

9.3   The Knapsack Problem

The Vytis Company has seven items $I_1, I_2, I_3, I_4, I_5, I_6, I_7$, that are to be shipped from New York to Detroit as quickly as possible. A company airplane with a freight capacity of 1500 pounds is available for this purpose. Table 9.3 gives the weight and

Table 9.3

	Weight (lb)	Value ($)
I_1	300	600
I_2	350	610
I_3	400	650
I_4	250	400
I_5	260	405
I_6	280	410
I_7	500	660

value of each of the seven items. Since the sum of the weights of these items is 2340 pounds and the freight capacity of the plane is 1500 pounds, not all of the items can be taken.

How should the plane be loaded so that the value of its contents is maximized and its freight capacity is not exceeded?

Corresponding to items $I_1, \ldots, I_7$, we introduce variables $X_1, \ldots, X_7$, respectively. X_1 can take on one of two values, 0 if item I_1 is not taken and 1 if item I_1 is taken; X_2 can take on one of two values, 0 if item I_2 is not taken and 1 if item I_2 is taken; and so on.

The function

$V = 600X_1 + 610X_2 + 650X_3 + 400X_4 + 405X_5 + 410X_6 + 660X_7$ is formed by multiplying X_1 by the value of item I_1, \$600, multiplying X_2 by the value of item I_2, \$610, and so on, then adding. V is the value function to be maximized. Whenever $X_1, \ldots, X_7$ are given values (0's and 1's), V becomes a sum of **assumed** item values. For example, when $X_1 = X_2 = X_3 = 1$ and $X_4 = X_5 = X_6 = X_7 = 0$, then

$$V = 600 + 610 + 650 = 1860,$$

the sum of the **assumed** values of I_1, I_2, and I_3.

The problem of loading the plane so that the **assumed** value of its contents is maximized and its freight capacity is not exceeded is expressed by the following integer program model:

Find nonnegative integers which maximize

$$V = 600X_1 + 610X_2 + 650X_3 + 400X_4 + 405X_5 + 410X_6 + 660X_7$$

subject to

$$X_1 \leq 1$$
$$X_2 \leq 1$$
$$X_3 \leq 1$$
$$X_4 \leq 1$$
$$X_5 \leq 1$$
$$X_6 \leq 1$$
$$X_7 \leq 1$$
$$300X_1 + 350X_2 + 400X_3 + 250X_4 + 260X_5 + 280X_6 + 500X_7 \leq 1500,$$

where $X_1, \ldots, X_7$ are defined in the preceding.

By inspection it is clear that the solution is $X_5 = X_7 = 1$ (take items I_5 and I_7), $X_1 = X_2 = X_3 = X_4 = X_6 = 0$ (do not take any of the other items). When the number of items is large, inspection is not a feasible approach to solving such a problem.

Here as well I emphasize to my classes that the solution obtained is **valid** with respect to the **assumed** weights and values of the items in question. Whether or not it's realistic to do so is another issue.

This problem illustrates problems with the following general structure. A container of some sort (truck, car, plane) is to be loaded with items of various **assumed** values and weights. For the items involved there is a limitation on the weight that can be loaded in the container but not on the volume. The problem is to load the container in such a way that its weight limit is not exceeded and the value of the items loaded is the largest possible. This is the immediate extension of the Vytis Company's Problem.

A problem with the same structure but very different appearance is illustrated by the Onutė Land Development Company's Problem posed in Section 9.4.

9.4   Food for Thought Questions

Formulate the following problems in terms of integer program models. State the postulates that underlie each model and the valid conclusion that follows.

1. The Vroman Institute has two jobs to fill, physiologist (job 1) and biochemist (job 2), and is considering three candidates, Ann (candidate 1), Gena (candidate 2), and Marty (candidate 3), for the jobs. Each candidate's qualifications for each of the jobs have been assigned a numerical rating. (see Table 9.4).

Table 9.4

		Physiologist (job 1)	Biochemist (job 2)
	Ann (candidate 1)	3	$\frac{5}{2}$
Candidate	Gena (candidate 2)	2	$\frac{5}{2}$
	Marty (candidate 3)	2	$\frac{3}{2}$

These ratings are interpreted as a measure of a candidate's potential for a particular job.

The Institute's problem is to fill the jobs in such a way that potential, expressed by the potential function, is maximized.

2. A legal advisory group is to make recommendations on two positions, State Supreme Court Judge and Civil Court Judge, and is considering three candidates, M. Jones, R. Johnson, and A. Marks. Table 9.5 describes potential ratings that have been assigned by the advisory group as a quantitative measure of each person's qualifications for each position.

Table 9.5

Candidate	Job	
	Supreme Court Judge	Civil Court Judge
Jones	9	8
Johnson	8	9
Marks	10	8

The advisory group wishes to make its recommendations based on how these positions should be filled on the basis of maximization of potential, expressed by the potential function.

3. The Onutė Land Development Company has identified five sites in Albuquerque, Dallas, Miami, Phoenix, and San Diego for the construction

of condominiums. The **anticipated (assumed)** cost of construction on these sites (in millions of dollars) and the **expected** profit to be realized from each development (in millions of dollars) are described in Table 9.6. The company can commit at most $32 million to these developments, which is insufficient to undertake them all.

The problem is to determine the selection of sites that yields the largest total **expected (assumed)** profit, but for which the total cost does not exceed the amount available.

Table 9.6

	Sites	Cost	Profit
(S_1)	Albuquerque	10	0.19
(S_2)	Dallas	12	0.23
(S_3)	Miami	11	0.20
(S_4)	Phoenix	15	0.30
(S_5)	San Diego	9	0.16

4. Ecap University has two jobs to fill, Dean of the Graduate School (job 1) and Dean of the School of Arts and Science (job 2), and is considering three candidates, J. Frank (candidate 1), M. Smith (candidate 2), and T. James (candidate 3). The search committee of the university, which is to make recommendations to the president, has assigned a numerical rating to each candidate's qualifications for each job (see Table 9.7), the search committee's **postulates**. These ratings are interpreted as a quantitative measure of each candidate's potential for each job.

The search committee wishes to make its recommendations in such a way that potential is maximized.

Table 9.7

Candidate	Job	
	Dean, Graduate School (job 1)	Dean, Arts and Science School (job 2)
Frank (candidate 1)	9	8
Smith (candidate 2)	7	7
James (candidate 3)	6	8

10

Index Number Modeling

10.1   Preface

Our first encounter with index numbers might leave us with the impression that there's nothing much to them beyond boring, boring computation. (That was my initial impression.) When the level is raised to modeling via index numbers our outlook is bound to change. The Consumer Price Index (CPI), to take the most important example of a real life index number, is often in the news because its value (the valid conclusion of an underlying CPI model) affects our lives in so many ways (sec. 10.4). Its nature and the controversy that surrounded it in recent years is the main focus of this chapter.

First, some observations about index numbers which might be useful background for further discussion

10.2   Index Numbers

Index numbers are used to measure the change in a quality that has occurred over time, or how a quantity compares with another. Index numbers are widely used to study fluctuation in business and economic activity. The initial time frame against which changes are measured is called the **base period** and the end of the time period is called the **given or current period**. Typically, the base and current periods are specific years and we will illustrate the various index number constructions using this format. However, index numbers are often calculated month by month, as well as year by year.

The value of the index at the base year is set at 100.0 so that the corresponding value of the index at any future time is either above, below or 100.0, depending on whether the given year measurement has increased, decreased or remained the same.

For instance, from January 1, 1996 to January 1, 1998 the Dow Jones Industrial Average rose from 5117 to 7908 or 2791/5117 = 54.5%. If base year (January 1 1996) aggregate stock prices are defined as 100.0, then aggregate January 1, 1998 prices equal 154.5. Between 1987 and 1989 the average price of fuel oil in New York declined by 8.9%. Thus, on average, the price of fuel oil in New York in 1989 is 91.1 relative to an initial value of 100 two years previous.

In comparison to the cost of living in Denver, Colorado in 2005, the cost of living in Milwaukee, Wisconsin in the same year was estimated as 6.6% higher, whereas the cost of living in the Tampa-St. Petersburg, Florida area was estimated as 11.1% lower. Consequently, we would assign a 2005 cost of living index of 106.6 for Milwaukee and 89.9 for Tampa-St. Petersburg, relative to the Denver, Colorado (base) index of 100, indicated by writing 2005 = 100.

James Company Stock

The average price per share of James Company stock was $14.23 in 2000, $17.37 in 2002, and $15.04 in 2004.

(a)   Write an index comparing the 2004 and 2002 average price per share of James Company stock to that of 2000.

(b)   Using 2002 = 100, determine an index that states the 2004 James Company's average price per share.

(a)   For 2004, $\dfrac{15.04}{14.23} \cdot 100 = 105.7$.

For 2002, $\dfrac{17.37}{14.23} \cdot 100 = 122.1$

Therefore, we can say that relative to the 2000 average price, the 2004 and 2002 average share prices of James Company stock were 5.7% and 22.1% higher, respectively.

(b)   For 2004, $\dfrac{15.04}{17.37} \cdot 100 = 86.6$.

Thus, the average James Company stock price was 13.4% lower in 1997 than it was in 2002.

Unweighted Index Numbers

Simple Aggregative Index

The simplest index number to construct is the simple or unweighted aggregative index, defined by

$$I = \frac{\sum P_n}{\sum P_o}, \qquad (1)$$

where $\sum P_n$ is the sum of the given-year prices and $\sum P_0$ is the sum of the base year prices. Their ratio is then multiplied by 100 to express the index as a percent.

Clothing Items of an Ecap University Student

Consider a collection of basic clothing items used by an Ecap University student, listed in Table 10.1. Construct a simple aggregative index comparing the 2008 prices of these items with their prices three years earlier.

Table 10.1

Item	Unit Price	
	2005	2008
Shirt/blouse	$15	$25
Socks (pair)	1	2
Jeans (pair)	23	38
Shoes (pair)	50	65
Sweater	16	25
Total:	105	155

Given two years of data for comparison, we generally consider the earlier year to be the base year and the later year to be the given or current year. Therefore, from the data the simple aggregative index for 2008 is given by:

$$I = \frac{\sum P_{2000}}{\sum P_{1997}} \cdot 100 \qquad (2)$$
$$= \frac{155}{105} \cdot 100 = 147.6$$

We can interpret this result in two ways. We can say that it cost 47.6% more in 2008 than in 2005 to purchase the group of items listed in Table 10.1. Equivalently, the 2008 aggregative cost of these goods is 147.6% of their previous value.

This procedure is simple to use, but has two major drawbacks. First, the index may be unduly affected by items having large price fluctuations. Suppose, for example, that Table 10.1 had listed a sixth article of clothing, a jacket whose price had dropped from $120 to $90 over the three year time span. Then the index would have been

$$\frac{245}{225} \cdot 100 = 108.9,$$

which suggests that there was only a small increase in overall prices, even though five of the six items in the group increased in price. The sharp change in the aggregative index was due to this sixth item having a much higher price in 2008 and 2005 than the other articles of clothing in the survey. Suppose, instead, that the sixth clothing item, underwear, diminished in price from $5 to $2, which represents a larger percentage drop in price (60% drop) than the previous change from $120 to $90 per jacket (25.0% drop), but which involves smaller dollar values. For this case the aggregative index would be:

$$\frac{157}{110} \cdot 100 = 142.7$$

This value is much closer to the original aggregative index, which is due to the use of smaller dollar values for the underwear and despite the steeper percentage drop in the price of the underwater compared to that for the jacket.

Another problem is that the aggregative index may be changed by altering the units of the items being compared; the more dramatic the unit change and the greater the number of commodities whose units are changed, the greater the potential change in the aggregative index. If in Table 10.1, for example, we had expressed the unit price of jeans in terms of every ten pairs, then the 2005 and 2008 prices per unit would have been listed as \$230 and \$380, respectively. Our aggregative index would then be:

$$I = \frac{25 + 2 + 380 + 65 + 25}{15 + 1 + 230 + 50 + 16} \cdot 100$$
$$= \frac{497}{312} \cdot 100 = 159.3$$

suggesting an overall price increase of 59.3% over the three year period in question, as opposed to the more modest estimate of a 47.6% increase obtained earlier for the same articles of clothing.

Thus, we see that the simple aggregative index may be easily, artificially and perhaps drastically altered to reflect different levels of price changes depending on the insights or whims of those conducting the analysis. For these reasons the simple aggregative index is not widely used today.

Arithmetic Mean of Price Relatives

We are able to eliminate the units problem, although not the high price fluctuation problem, by employing price relatives. For each commodity we compute the ratio of its price for the given year to that of the base year. The value of this ratio is fixed, irrespective of the unit used for the commodity. Then we use any measure of central tendency to arrive at a "typical" ratio. The arithmetic mean is generally used for this purpose. On carrying out these steps we obtain the arithmetic mean of price relatives, defined by

$$I = \frac{\sum \frac{P_n}{P_o} \cdot 100}{k} \tag{3}$$

where k is the number of items whose price relatives are being averaged to form the index.

A Return to the Clothing Items Situation

Based on the data of Table 10.1, construct an arithmetic mean of price relatives measuring the overall change in the prices of the given articles of clothing from 2005-2008. Use 2005 = 100.

After dividing the 2008 price of each item by its 2005 price and multiplying by 100, we obtain Table 10.2,

Table 10.2

Clothing	Price Relative
Shirt/blouse	$\frac{25}{15} \cdot 100 = 166.7$
Socks (pair)	$\frac{2}{1} \cdot 100 = 200.0$
Jeans (pair)	$\frac{38}{23} \cdot 100 = 165.2$
Shoes (pair)	$\frac{65}{50} \cdot 100 = 130.0$
Sweater	$\frac{25}{16} \cdot 100 = 156.3$

Total 818.2

Accordingly, we find the arithmetic mean of price relatives index to be:

$$I = \frac{818.2}{5} = 163.6 \qquad (4)$$

We interpret this result as meaning that prices have increased 63.6% on average for the clothing items over the course of the three year period. Not surprisingly, this result differs from that obtained by using the simple aggregative index.

As before, if one or few commodities undergo significant price changes, this may significantly alter the value of the mean of the price relatives index. For instance, suppose that the price of a pair of shoelaces tripled from $0.50 to $1.50 between 2005 and 2008. If we include this accessory item in our price relatives computation, the index becomes

$$I = \frac{818.2 + 300}{6} = 186.4,$$

which is significantly larger than 163.6, the arithmetic mean of price relatives index obtained from (3) without the shoelaces.

On the other hand, if we include the shoelaces, the revised simple aggregative index becomes

$$I = \frac{156.5}{105.5} \cdot 100 = 148.3,$$

which differs little from 147.6, the simple aggregative index obtained from (1) prior to inclusion of the shoelaces.

Many would consider the change in the latter index to be the more realistic in this case. Even tripling the cost of a pair of shoelaces should have a minimal effect on the overall change in the price of clothing because its cost is still low relative to that of other articles.

The addition of this low cost item into the price relatives calculation had a disproportionate effect on the index. If we took into account the quantities used of these articles, we might get a more realistic picture of overall price changes.

Unweighted index numbers are straightforward to compute. However, aside from the drawbacks we have noted, a major disadvantage of the unweighted price index is that it makes no distinction between high and low turnover items, so that a large price change in even one or two low turnover commodities could significantly distort the index. Suppose, for example, that we add a raincoat to our previous list of clothing items and its price dropped from $105 to $50 between 2005 and 2008. From (1) and Table 10.1 and (3) and Table 10.2, this would yield the following revised values of the aggregative and price relatives indexes, respectively:

$$I = \frac{155+50}{105+105} \; 100 = 97.6$$

$$I = \frac{818.2 + \dfrac{50}{105}\,100}{6} = 144.3$$

The revised price relative value of 144.3 suggests an average price increase of 44.3%, rather than the previous one of 63.6% (see (4)). The revised aggregative index of 97.6 suggests that, on the whole, prices have actually declined by 2.4%, rather than the previous one which suggests a 47.6% (see (2)) overall price increase!

Over the course of a year a "typical" student would purchase the other articles more often than a raincoat, and a more realistically designed price index would have hardly changed despite the 52% drop in the price of the raincoat. Furthermore, even a relatively small price change in a high usage item may have a significant impact which is not reflected by unweighted indexes.

Because of the ways in which unweighted indexes may give misleading results, they are not commonly used today in important price analyses. Prior to 1914 the Wholesale Price Index of the Bureau of Labor Statistics was computed as an arithmetic mean of price relatives of about 250 commodities. Around this time it was changed to a weighted index.

10.3   Index Number Models

An index number, as the afore examples illustrate, does not stand alone. It has a family in that it is a valid consequence of a number of **assumptions** about the items chosen, the years chosen for comparison, regions chosen as a source of data, and the mathematical method used to make the comparison; unweighted aggregative method vs. unweighted arithmetic mean of price relatives method vs. weighted aggregative method vs . . .

In his treatise *The Making of Index Numbers* (1922) Irving Fisher describes 134 approaches to constructing weighted index numbers. Each such structure together with the background and underlying **assumptions** made, and the index number obtained as a valid conclusion is called an **index number model**.

10.4   The Consumer Price Index

The Consumer Price Index (CPI) is one of the most generally accepted ways to measure the behavior of inflation. It charts the average change in prices over time of a fixed market basket of goods and services. The Bureau of Labor Statistics releases this information for two different population groups: (1) A CPI for All Urban Consumers (CPI-U), which covers about 80 percent of the total population, and (2) A CPI for Urban Wage Earners and Clerical Workers (CPI-W), which covers about 32 percent of the total population. Of course, in light of these percentages, there is some overlap in the groups covered by these indexes.

The CPI notes changes in the price of such items as food, clothing, housing, energy, transportation, medical and dental services, medical drugs, and other goods and services that people require for day-to-day living. Different urban areas, housing units and retail establishments around the country are taken into account in compiling this index. Then price changes for the various categories in each region are weighted to take into account the relative amount spent by that particular locale. Finally, local data is combined to obtain an overall average.

10.5   The CPI and Us

Determining "Real" Dollar Value

As we have all experienced, what a dollar bought yesterday is not what it buys today. What it buys today is not what it can be expected to buy tomorrow. If comparisons of a dollar's worth over time are to be meaningful, we must have a mechanism for adjusting dollar values to reflect "real buying power" compared to some suitable base used as a point of reference. Price indexes provide such a mechanism. The process of using price indexes to make these dollar value adjustments is called **deflation** and the price index used as a divisor is called a **deflator**. The dollar figure that results from this deflation procedure represents "constant dollars" or "real wages"—that is, the purchasing power of the obtained (deflated) dollar value in the base year. Formally expressed, we have:

$$\textbf{Real Dollar Value} = \frac{\text{Current Dollar Value}}{\text{Price Index}} \cdot 100$$

"Real" Gross Domestic Product?

Table 10.3 shows unadjusted values of the Gross Domestic Product (GDP), in billions of dollars, for the United States in 1992 and 1995 as well as the GDP price deflators (aggregative price index numbers) for these years, in terms of 1990. Calculate their percentage growth in actual (unadjusted) GDP. Determine the real GDP in 1990 prices and the percentage growth in real GDP.

Table 10.3

	Year	
	1992	1995
GDP	6244.4	7253.8
Price Deflator (1990 = 100)	106.8	114.9

Percentage growth in actual GDP between 1992 and 1995 is:

$$\frac{7253.8 - 6244.4}{6244.4} \cdot 100 = 0.162 \text{, which translates to } 16.2\%$$

Real GDP in 1990 prices for the years 1992 and 1995, respectively, are:

$$\frac{6244.4}{106.8} \cdot 100 = \$5846.82 \text{, } \frac{7253.8}{114.9} \cdot 100 = \$6313.14$$

Therefore, percentage growth in real GDP during 1992-1995 is:

$$\frac{6313.14 - 5846.85}{5846.82} \cdot 100 = 0.08 \text{, which translates to } 8.0\%$$

We may conclude from this analysis that, although actual GDP grew 16.2% in this time span, only 49.4% of this increase (namely 8.0/16.2 times 100) may be attributable to real growth in GDP, whereas the remaining 50.6% of the actual growth is attributable to inflation.

It all looks so *precise* and it is, in the math sense that the computations were correctly carried out. But this does not mean that GDP is the most appropriate measure of national income, of how much wealth Americans make, I point out to my students. It's a tough sell because of the tendency to equate mathematical *precision* with real-world relevance of the (valid) conclusions reached to the situation under study.

For discussion of this particular matter see Eric Zencey, "G.D.P. R.I.P.", *The New York Times*, Aug. 10, 2009, A15, and David Vanderpool, "G.D.P. Flaw: Not All Economic Activity is Productive", (letter in response to the afore) *The New York Times*, Aug. 17, 2009. For general discussion of the relevance of numbers to a situation under study see ch. 22.

The CPI's Wide Reach

The CPI's wide reach is well-described in an article of that title by John M. Berry [6].

Taxes

Taxes are adjusted in a variety of ways to protect taxpayers from the effects of inflation. Some of the features adjusted each year in line with the CPI are:

- **Tax bracket break points,** which determine the rates imposed on income at different levels.

- **Personal exemption and standard deduction,** including the additional standard deduction for the aged and the blind.

- **Earned Income Tax Credit,** the maximum amount of credit and the income range over which the maximum credit is phased out.

- **Limit on itemized deductions**

- **Pension contribution limits**

- **Excess pension distribution tax**

Fully Indexed Programs

Fully indexed programs are those in which automatic increases in benefits levels or eligibility are directly determined by the CPI. A smaller CPI change would mean benefits would not rise by as much in the future. These programs include:

- **Social Security,** with nearly 45 million beneficiaries, including retirees and disabled workers, their spouses and eligible children including survivors of deceased workers.

- **Railroad retirement,** with around 800,000 beneficiaries, including retirees, their spouses and eligible children.

- **Supplemental Security Income,** federal welfare for 6.5 million elderly and disabled Americans.

- **Veterans' compensation,** benefits paid to people with disabilities linked to service, and veterans pensions.

- **Federal military and civilian employee pensions,** paid to 4 million retirees.

- **The official poverty line,** by definition, rises each year in line with the CPI. Eligibility for numerous federal programs is determined by comparing income with the poverty line. Affected are 26 million recipients of food stamps, 25 million in subsidized child nutrition programs, more than 5 million with federal student grants and about the same number getting energy assistance.

Elsewhere

The CPI also is used in many other ways to adjust for inflation:

- **Rents.** Some contracts, particularly for commercial space, include escalator clauses using the CPI. Some localities with rent control laws use the CPI to determine the maximum allowable rent increases.

- **Employee compensation.** Some state and local government pay and pensions, some labor contracts and some company compensation schemes include cost-of-living-adjustments, or COLAs, pegged to—although not necessarily equal to—changes in the CPI.

- **Alimony and child support payments.** Some divorce agreements include escalator clauses linked to the CPI.

10.6   Limitations of Index Number Models

Beware the Assumptions

It is important to always keep in mind that index numbers are valid conclusions of **assumptions** of mathematical models of our making to help us describe and predict the movement of highly complex phenomena.

Sometimes these valid conclusions do the job well and sometimes they miss reality's mark by a wide margin. The more complex the phenomenon, the more difficult it is to realistically capture its essence in terms of a math model and predict its behavior terms of index numbers.

10.7   How *Accurate* is the CPI?

This question follows from the afore observations about the limitations of index number models; the CPI, after all, is the super-star of all index numbers in terms of scope and complexity. It is also clear that the real world accuracy of the CPI as a measure of inflation is of paramount importance.

During a period in the 1990's when the overriding focus was on achieving a balanced national budget the thrust that emerged was to narrow the budget gap by lowering the value of the CPI. The following observations by Louis Uchitelle [29] and Dean Baker [3] provide perspective on the delicate technical balancing act and the political atmosphere that confronted the Bureau of Labor Statistics that was charged with developing the CPI model.

L. Uchitelle; Balancing Quantity, Quality and Inflation

As heart bypass operations are perfected, more people have them, extending their lives. But the operations raise health care costs. Is this inflation, as the Government's statisticians contend? Or is it simply fair value for the promise of longer, healthier lives, as a blue-ribbon panel of economists argues?

Or consider what happens when people shift from going out to the movies to bringing videos home. Watching "Independence Day" at the neighborhood theater costs, say, $7 a ticket while the video rental is $3.50, no matter how many people crowd around the television. Both prices have held steady for a while. Is that zero inflation, as the Government says? Or should the Consumer Price Index register a price decline for movies as people shift to a different movie experience that is less costly but comparable in quality, as the economists contend?

In the suddenly expanding debate over the accuracy of the C.P.I.— which is used to set everything from Social Security pensions and union wage increases to individual tax brackets—questions like these are crucial. Indeed, behind the controversy, which was set off when the Congressionally appointed panel of economists recently released its report, lie two fundamentally different ways of looking at the economy.

The blue-ribbon panel, led by Michael A. Boskin, President George Bush's chief economic adviser, concluded that the C.P.I. exaggerates the rate of inflation by slightly more than one percentage point a year. Defenders of the current way of measuring price changes—a group that includes representatives of the elderly, labor unions and a number of economists as well—contend that any error is much smaller.

If little were at stake, most people would be happy to leave it to the experts to argue over a few tenths of a percentage point one way or another. But much is at stake—each tenth adds or subtracts billions of dollars in Government payments now pegged to the C.P.I.—and so a struggle has developed for control of the Consumer Price Index.

Today, health care and movies are among the thousands of items purchased by Americans that are cranked into the calculation of America's standard inflation gauge. The C.P.I. essentially measures changes in the prices of a relatively static market basket of goods and services.

But if the measure of inflation is redefined to put more emphasis on what it takes to maintain a given standard of living, then subjective value judgments must play a much bigger role. Higher costs for open-heart surgery, instead of registering as inflation, might instead be counted as a qualitative improvement in the chances of living a better life.

Video rental and theater tickets are listed separately in the C.P.I. But with VCR's in so many homes now, a video today might be judged a reasonably satisfying alternative to a night at the movies. Shouldn't that be reflected somehow as a lower cost to purchase the same living standard?

So the debate intensifies, and as it does, very different views emerge on measuring human well-being.

"I don't think we know how to construct a cost-of-living index that would deal with all of the issues about quality of life to everyone's satisfaction," said Katherine Abraham, Commissioner of the Bureau of Labor Statistics.

The bureau, an independent agency within the Labor Department, is the keeper of the Consumer Price Index, calculating each month the price changes for a market basket of 71,000 items. But now the bureau is under pressure from the commission of five economists, appointed by the Senate Finance Committee, to recast the index so that the C.P.I. does more to quantify the values that people attach to what they buy.

"Such values are not chiseled in stone," said Zvi Griliches, a Harvard economist and a commission member. "They are squishy numbers, but they are better than a firm but wrong zero."

Adopting the commission's outlook would change the economic picture drawn by the Consumer Price Index, making American living standards look more solid. The C.P.I. is currently rising at an annual rate of 3.2 percent, but if the panel's estimates are accurate, the annual rate is really closer to 2.1 percent.

The report is already serving as a catalyst for action. With commission members setting the pace, discussions are under way among economists across the country in an attempt to develop a consensus behind reducing the growth rate of the C.P.I. by late next year. Rather than count on the Bureau of Labor Statistics to adjust its numbers, there is a move afoot, pushed by the Clinton Administration, for Congress to act.

"The bureau is a very important expert in this process," Treasury Secretary Robert E. Rubin said, "but there are a lot of other experts as well. Congress, in acting, would have to reflect a broad-based agreement among these experts about changes in the C.P.I. that would cause it to better reflect inflation."

If such a consensus emerges, the Administration might back a bill that would leave the C.P.I. unchanged, but—for purposes of determining Federal benefits and tax brackets—would subtract four- or five-tenths of a percentage point from each published index figure. The billions of dollars saved would reduce the budget deficit, a goal of both major political parties.

But complicated value judgments must be incorporated into the Consumer Price Index to slice off that initial fraction. Mr. Boskin, now a Stanford professor, argues that the bureau could reach the goal "in the time frame of a year," by "expanding, improving and accelerating" studies it already has under way.

Ms. Abraham is not so sanguine. Although she accepts many of the commission's proposals in principle, she insists on moving cautiously. "There are a lot of issues associated with the measurement of the cost of living that I don't think we will solve in my lifetime," said Ms. Abraham, who is 42.

Her bureau is constantly studying changes, she says, that might make the C.P.I. more of a measure of the cost of maintaining one's standard of living. Changes adopted in the last two years have already reduced the level of the C.P.I. by 0.22 percentage point, she notes. And the bureau is considering others that the Boskin commission would applaud.

The huge market basket of goods on which the C.P.I. is based is to be updated more frequently to reflect shifts in the purchases that Americans actually make. And new products are to be added more quickly. The cell phone is one. It is still not calculated in the national index, although its falling price might dampen the inflation rate. "That was a mistake, not introducing cell phones sooner," Ms. Abraham said.

Still, in other ways, she and her lieutenants are digging in their heels. Shifting the C.P.I. to a standard-of-living index is not the goal, they say, accuracy is, and that rules out what some officials describe as "hip pocket guesswork" in trying to quantify how people value the goods and services they buy.

How, for example, do they feel about renting videos instead of going to the movies? Mr. Griliches, the Harvard economist, acknowledges that the two experiences are not interchangeable for everyone, but they are for enough people to drive down the price of maintaining this particular pleasure.

"I would not argue that the cost fell from the $7 ticket price to the $3.50 movie rental," Mr. Griliches said, "but it wasn't zero either. We are trying to make an estimate and this is where it goes ad hoc. I would try to base it on how many people shifted from one to the other. Or I would do a crude approximation and say half the difference in price is a real decline in the cost of experiences that are similar in quality."

But Patrick Jackman, chief of the division at the bureau that calculates each month's C.P.I., is not swayed. "You can get off cheaper by viewing the movie at home," he said, "but you may not get the same pleasure of an evening out away from the kids in a movie house, which can be a communal experience."

The commission and the bureau differ, too, over Zantac, the heart-burn medicine. It is rapidly replacing Tagamet, which many doctors consider less effective. The bureau attributes the higher cost, about 60 cents a pill, to higher quality, not inflation. But the commission would go further, counting it as a price decline on the ground that the quality improvement exceeds the price increase.

Similarly, the bureau undervalues quality improvements in autos, says Robert Gordon, a Northwestern University economist who was also on the Boskin commission. Too many of the price increases in autos have been attributed to inflation, he says, rather than taking into account that cars are more durable and require less maintenance.

The classic example in this debate is meat versus chicken. Meat prices rise, so people shift to less expensive but nutritionally equal chicken. The C.P.I. registers an increase in inflation, because of the higher meat prices. But the panel calls this a decline in inflation: less cost to maintain the same standard of living. "The Consumer Price Index pretends that people are not free to shift to chicken when that is cheaper, or sweatshirts when sport shirts are too expensive," Mr. Gordon said.

But in judging these shifts, isn't the Boskin commission being too optimistic, often failing to recognize deteriorations in the quality of American life, too? That is the view of Dean Baker, an economist at the Economic Policy Institute. "If people perceive they are making a switch in response to a rising price," he said, "they can perceive a loss in the quality of their lives that may not be fully picked up by the Boskin commission's recommended price index."

That complicates the puzzle, and the commission recognizes this. To quantify the most difficult value judgments—representing roughly six-tenths of a percentage point of the 1.1 points that the commission would shave off the C.P.I.—the panel would establish a separate, parallel index. The second index would deal with such things as crime-fighting, where value judgments require measurements over a period of time.

"If I install a better alarm system," Mr. Griliches said, "that is an improvement in the quality of my life, and therefore a decline in inflation. But if the burglars learn how to trick this alarm system, that is a rise in price, because the quality advantage will be eroded. So one of the issues is that this whole notion of quality, of standard of living, is not a constant."

Such complexities give Ms. Abraham pause. But the panel contends that properly handled, they can be incorporated relatively soon, by means of the second index.

This parallel index would be adjusted, up or down, in hindsight over a one—or two-year period, to deepen understanding of living standards and the economy. The regular C.P.I., tracking each month's changes, cannot be adjusted; too many payments are tied to it.

A REPORT by economists on the shortcomings of statistics would not normally send politicians' pulses racing. But when it promises to cut billions of dollars from the budget deficit, statistical detail becomes big politics. The findings of a commission set up by the Senate Finance Committee to study the calculation of America's consumer price index (CPI), released this week, do just that. A group of five eminent academics led by Michael Boskin, a professor at Stanford University, has concluded that the CPI probably exaggerates true increases in the cost of living by something like 1.1% a year, although the exact figure could be anywhere from 0.8% to 1.6%.

A correction of this magnitude would have enormous implications for the budget. Around a third of federal spending, mostly in retirement programmes, is directly indexed to changes in consumer prices. Through the indexing of individual income-tax brackets, a change in the CPI affects federal revenues, too. The Boskin commission reckons that correcting the 1.1% overstatement would save the government around $1 trillion over the next 12 years. By 2002, the target year for balancing the budget, around $200 billion could have been saved. Small wonder, then, that many politicians hope to prune public spending by simply improving statistics.

Among economists, the notion that inflation may be mismeasured is nothing new. Zvi Griliches and Dale Jorgenson of Harvard University and Robert Gordon of Northwestern University—all members of the commission—have spent years estimating the size and cause of the bias. Nor does the impact of mismeasurement affect the budget alone. Revising the CPI would substantially raise official measures of America's growth rate and hence of Americans' economic prosperity.

What are the problems? The basic difficulty is that the CPI is not really a measure of changes in the whole cost of living, but rather a gauge of price rises in a fixed basket of goods. This means that the CPI does not keep up with changes in the products consumers buy, or in how and where they buy them. According to the Boskin report, about a third of the mismeasurement (0.4%) is due to the fact that the official price index fails to capture important changes in consumer spending patterns. If the price of hardback books increases, for instance, people may buy more paperbacks. Or if bananas become more expensive, they may switch to apples. These switches do not make consumers feel worse off; but the CPI records a price increase.

A further 0.1% bias is due to the fact that Americans now do more shopping at discount outlets. But the biggest effect (0.6%) comes because the CPI underestimates the benefits shoppers gain from improvements in product quality and from the plethora of new products. Cellular telephones, for instance, are not yet included in the CPI, although 40m Americans now own one.

The Boskin report makes a host of suggestions as to how some of these inaccuracies could be corrected. The main message is that the Bureau of Labour Statistics (BLS), the federal agency that works out the CPI, should explicitly try to build a cost-of-living measure. No one is exactly sure how big an improvement such a change would make, although the commission's boffins reckon 0.4% a year could be achieved fairly quickly. That would bring budget savings of around $80 billion by 2002.

But the BLS, a fiercely independent if rather stodgy outfit, may not want to rush into dramatic changes. It is already trying hard to improve the numbers: a correction earlier this year shaved about 0.2% from the CPI. When the system's weights are next updated in January 1998, a further 0.1-0.2% may be won, although the updated index will still reflect only the buying patterns of 1993. And the BLS is introducing a new index, somewhat similar to the one the Boskin commission proposes, on an experimental basis early next year, although the agency's economists are quick to point out that the new index has drawbacks too.

Assume, for a moment, that the BLS is prodded into introducing such changes. Cost-of-living rises will still be overstated. And, although the report has some ideas about how to account better for quality improvements, they are likely to pare at most 0.2% from the CPI over the medium term. The problem is that the economy is changing too quickly for the statisticians to keep up. However much the statistics are improved, the CPI will still overstate true increases in the cost of living.

This leads to the bigger political point. If the CPI, however revised, is not an accurate guide to changes in the cost of living, should it be the index to which retirement benefits are linked? The Boskin report makes the point explicitly: Congress and the president, it suggests, must decide whether they wish to continue widespread, substantial over-indexing of benefits. In other words, if politicians are looking for budget savings, this report gives them intellectual backing for proposing that retirement benefits should be indexed to something less than the CPI.

Dean Baker:
The Inflated Case Against the CPI

There is now the appearance of an expert consensus that the government's most important measure of inflation, the consumer price index (CPI), seriously overstates the true increase in the cost of living. This sudden enlightenment is less the result of new research than political convenience. A cut in the CPI would reduce government payouts and ease the path to deficit reduction. Even better, it would do so via a technical adjustment that left few political fingerprints.

Tax brackets and government benefit programs such as Social Security are indexed to the CPI. If the CPI overstates inflation by 1 percent, as the Senate Finance Committee's Boskin panel has proposed, and the index is adjusted accordingly, this would reduce benefits and the deficit by a cumulative total of $634 billion over 10 years. Not bad for a technical fix.

Doubtless, the way we measure inflation requires continuous refinement. The Bureau of Labor Statistics (BLS) takes this task seriously, and has made myriad small adjustments over the past three decades. For years, there was a nuanced and relatively obscure debate about how to fine-tune the CPI. Lately, there has been a politically driven frenzy, as a small group of economists has labored to uncover—and exaggerate—all the ways in which the CPI might overstate inflation. Many claims have been advanced based on very little real evidence. There has also been virtually no effort to examine the ways in which the CPI might understate inflation.

The immediate protagonists are the five-member panel appointed by the Senate Finance Committee, chaired by former Bush economic advisor Michael Boskin, to make recommendations on revisions in the CPI. Though the group includes some eminent economists, all had previously testified on the CPI's supposed bias. All were chosen as known quantities who could be reliably counted upon to recommend a downward revision. Other eminent economists such as former BLS Commissioner Janet Norwood, who took the opposite view, were ignored. The panel was appointed in June 1995 and announced its 1 percent solution in mid-September. It conducted no original research. Instead, it used rough rules of thumb to reach its conclusions.

The Case For Shrinking The CPI

Five factors are usually cited by those claiming an upward bias in the CPI. Each provides some basis for claiming the index overstates inflation. However, the size of any resulting overstatement is far smaller than what is being claimed, and may well be offset by the sources of understatement in the CPI.

Substitution Effects. This is the most frequently cited source of bias, perhaps because so many reporters learned about it in their introductory economics classes. Most goods have close substitutes. If oranges are $1.99 a pound, consumers switch to apples. The CPI measures the prices of a fixed basket of goods and services. When the price of some goods in this basket rises temporarily, thrifty consumers shift to substitutes. By holding the basket fixed, the CPI then overstates the true increase in the cost of living for most consumers. This is a fair criticism.

However, most studies that have tried to measure the size of this bias find it to be very small, between 0.1 percent and 0.2 percent annually. (The Boskin panel scored it as 0.3 percent.) Moreover, one might fairly argue that even a close substitution "choice" dictated by a price rise entails an offsetting loss to quality. Presuming the substitute to be identical is like comparing, well, apples and oranges.

The Wal-Mart Effect. Over the last several decades discount stores have displaced many traditional retailers. As a result, consumers purchase many goods at far lower prices. The CPI does base its local samples on where consumers actually shop, but the CPI does not record the switch from a traditional department store to a discount store as a price decline. Rather, the price differential is treated as offsetting the lower quality of service in the discount store. Clearly this treatment misses a cost saving. Since discount stores have grown rapidly at the expense of traditional retailers, many consumers must consider the cheaper price well worth the lower quality service.

However, the importance of this difference for the CPI has been vastly overstated. Only about 15 percent of the index consists of goods that could potentially be sold in discount stores (primarily apparel, appliances, and household furniture). The share of consumers who patronize discounters versus full-price retailers simply does not change much from year to year. Moreover, even if the true price difference is as much as 10 percent (after adjusting for differences in service quality), this would lead to a bias of just 0.015 percent a year. This compares to a figure of 0.2 percent to 0.4 percent often cited by those claiming a substantial CPI overstatement of inflation. The Boskin panel used a figure of 0.2 percent.

Quality Bias. One of the largest sources of alleged overstatement of inflation is the failure of the consumer price index to fully account for the improvements in product quality. Clearly, a $2,000 computer today is a far superior machine to a $2,000 computer bought as recently as 1994. Most products are continuously improving in quality. But in fact, the consumer price index already includes extensive adjustments for product quality. It may even overstate the improvement in quality in some cases.

For example, the price index for new cars has increased approximately 150 percent from 1970 to the present, although the average price of an actual new car has increased about fourfold. Based on the CPI's price adjustment, it should be possible to purchase a car today for approximately 2.5 times what a comparable car cost in 1970. In 1970, a new Volkswagen Beetle cost about $2,000. But no new car is on the market today for anything close to $5,000, as would be implied by the new car index. The bottom-of-the-line new car available today for $9,000 is doubtless a significantly better car than a Volkswagen Beetle, but there are probably many consumers who would prefer to purchase a new Beetle at $5,000 rather than pay more for a better car.

This is the general problem with the quality adjustments in the index. Price increases attributed to quality adjustments are not counted as price increases in the index, even though many consumers might not pay for them if they had the choice.

There are also many areas where quality has plainly deteriorated. For example, many consumers have to spend more time fighting with health insurance companies over the processing of claims than 20 years ago, but this deterioration is not picked up in the index. Airline service has declined—less legroom, more changes of planes and missed connections, fewer meals, more convoluted fares. Other examples include longer waits in traffic and time spent waiting on hold or navigating voicemail instructions. None of these quality deteriorations are recorded in the CPI. Given the very limited research on quality adjustments in the CPI (the most frequently cited work is now twelve years out of date), it is impossible to reach any conclusive judgment about the size, or direction, of quality bias in the present index.

New Products. New goods are typically not included in the index until several years after they first appear on the market. During this time, they often undergo large price reductions—which are not picked up in the index. The inclusion of these price declines would lead to a lower measure of inflation. The classic example of this problem is the hand calculator. When it first appeared on the market it cost over $1,000. Within a couple of years its price had fallen to under $100. This huge decline in price was not picked up in the CPI.

But wait. Most people don't buy very expensive new products. Such products only find mass markets after their price has come down dramatically. (How many people bought thousand-dollar hand calculators?)

The CPI is an "expenditure-weighted index," meaning that each dollar of consumer expenditure is weighted equally. This means that if Donald Trump spends 1,000 times what the average consumer spends, his expenditures count 1,000 times as much as those of the average consumer. Of course, it is the Donald Trumps who buy the expensive new products.

However, it is possible to construct the index in a different way, which would count each consumer's expenditures equally. Under this system 1 percent of my budget would count the same as 1 percent of Donald Trump's budget in determining the weighting of a particular item. Such an approach is clearly more appropriate for the purposes the CPI is intended for. There is no reason that Social Security recipients should get a smaller cost of living adjustment because a few wealthy people experience huge savings on hand calculators. Nor does it make sense that such a bonanza for the wealthy should affect wage contracts that use the CPI as a point of reference.

A "person-weighted" index that counted each individual's expenditures equally would provide a much better gauge of the increase in the cost of living experienced by most of the population—and it would virtually eliminate the problem of new goods as a source of bias in the index. Here is a case crying out for a genuine technical adjustment, but one that cuts in the opposite direction from the Boskin panel.

Formula Bias. This is the most technical issue (sorry). Suppose the price of a good rises from $1.00 to $1.10. This is a 10 percent increase in price. Suppose it then falls back from $1.10 to $1.00; this is a 9 percent decrease in price. If these changes were just added together it would imply a 1 percent increase in price even though the price had not changed at all. BLS was never so foolish as to construct the CPI in such a way that it would be generally subject to this bias. But there are other such technical problems. BLS has researched this issue extensively, and uncovered and corrected several such areas. It is possible that problems of this sort may still exist in places, but the impact is likely to be extremely small, almost certainly less than 0.1 percent annually.

Let's give the critics the benefit of the doubt. Adding all of these possible biases together, a plausible estimate of inflation overstatement in the CPI would be 0.4 percent—not the one full percentage point estimated by the Boskin panel. But this is a gross adjustment, not a net one: It revises only those elements of the CPI that apparently overstate inflation. To be accurate, one needs to offset this adjustment with factors that suggest the CPI may be too low.

A Downward Bias?

The possibility that the CPI understates inflation has received little attention, since it won't help cut the deficit. Indeed, if the CPI errs on the low side, then Social Security pensioners should be getting bigger checks. Consider three distinct sources of downward bias:

Health Insurance Costs. The CPI does not include most increases in insurance premiums for individual health insurance, or increases in copayments or deductibles on employer-purchased insurance. This has been a major drain on household budgets in recent years, as employers have shifted more of the cost of health insurance back to their employees. Nor does the CPI include increases in required payments by beneficiaries of government programs such as the proposed increases in the Medicare Part B premium. This exclusion would lead to an enormous understatement in the cost of living of an elderly household. Even under the Clinton Medicare proposal, the premium is scheduled to rise by approximately $500 per beneficiary over the next seven years. This increase will consume nearly 5 percent of the annual income of a couple with an income of $20,000, the midpoint of the income distribution for families over 65. Under the Gingrich proposal, the total increased costs to consumers would be substantially higher.

Today's basket of health services is in many respects superior to that of, say, two decades ago—thanks to new technologies, drugs, and lifesaving procedures. It's also true that under managed care, many doctors are more harried and patients are often rushed out of hospitals. Some of the higher cost of today's health care reflects not better service, but deadweight losses—the cost of claims processing, risk selection, mergers, and acquisitions. Spending on health care has quintupled in three decades, but it's not at all clear that the quality has. It would take a great deal more research to determine how to net out the improvements and degradations to quality, and then weigh them against the unambiguously higher cost.

Personal Business Expenditures. This category of expenditures, which includes items such as lawyers' and brokerage fees, has been rising at the rate of 0.2 percent a year for the past 20 years as a share of disposable income. This is a category of spending that may not provide direct benefits to consumers but rather is often a cost of maintaining a standard of living threatened by deteriorating external circumstances. Suppose, for example, that I have to hire a lawyer in order to resolve a dispute with my health insurance company, because health insurers are becoming more aggressive. Compared to a situation where the insurance company deals with me honorably, this is a needless expense. I am worse off in direct proportion to the amount of money that I have to spend on my lawyer, regardless of the quantity or quality of legal services provided.

Similarly, if I have to rely on a private financial counselor to ensure a secure retirement, instead of receiving a company pension, I am worse off in direct proportion to the amount of money I must pay the counselor, not the fee per financial transaction. Likewise the cost of divorce lawyers and security consultants. Since the CPI only measures the change in the price of these services, rather than the change in the overall need for the services or genuine benefit derived, it is understating the true increase in the cost of living.

Quality of Life Factors. Many factors that affect the quality of life are not picked up in the CPI. An obvious example is crime. If people have to spend more money in order to live in a safe neighborhood—say on security alarms—it is not captured by the CPI. Nor is a deterioration in the quality of public schools and therefore an increased need for spending on private schools or personal tutors. Nor is the cost of joining a private exercise club because the local public facility has closed or become unsafe. Many of these issues are quite complex and cannot be easily quantified in any meaningful way, but this doesn't make them any less real. Anyone attempting to make conclusive statements about changes in the "true" cost of living must be prepared to address such issues.

Given the very limited amount of research on the biases in the CPI, it is impossible at this point to reach any conclusive judgment about the magnitude or direction of the overall bias in the index. While the claim that "the CPI overstates inflation" has become a virtual mantra of the Washington punditry, honest proponents of this view acknowledge that it is based on very little evidence. The best solution would be to provide the professional statisticians at BLS with the resources they need to improve the CPI. This is clearly preferable to bending economic statistics in whatever direction is politically expedient.

Unexpected Twists

One striking thing about this whole debate is that conservatives try to have the argument both ways. Supposedly, a lower CPI would cut the deficit; and cutting the deficit would stop us from "robbing our grandchildren." But if the Boskin panel is right and the CPI is overstated by 1 percent per year, then real median wages are rising at 2 percent per year. This means that they will double in approximately 35 years, so that the average real wage will be approximately $50,000 a year (in 1995 dollars) in the year 2030. If so, our grandchildren will do just fine and there's no need to slash the Social Security of their grandparents.

An overstated CPI also means that people had been much poorer in the recent past than we realized. If the CPI was overstated by 1.5 percent in the past, as suggested by the Boskin Commission, then the average annual wage in 1960 was just $11,215 measured in 1995 dollars. In 1960, today's 70-year-olds were 35. It is hard to justify taking Social Security benefits from these people in order to make the 35-year-olds of 2030 better off.

In addition, the critics have ignored the implications of a significant CPI revision for monetary policy and growth. Alan Greenspan, chairman of the Federal Reserve, has testified that the CPI overstates inflation by as much as 1.5 percent. But if the official inflation rate is lowered, it undermines the entire rationale of Greenspan's tight money policy. So politicizing the CPI leads to some unexpected implications. It would be far better to return the question of CPI revision to intellectually honest technicians, where it belongs, and to argue deficit reduction on its merits.

10.8   Politics vs. Mathematics

Ideology vs. Mathematics in the Soviet scene in the 1940's and early 50's (ch 8; sec. 8.9) is one example of the intrusion of the world scene into mathematical analysis. Another, the American version, is Politics vs. Mathematics concerning the Consumer Price Index in the middle to late 1990's.

In 1995 center stage of government business was taken up by the problem of eliminating the federal budget deficit. As part of their plan to accomplish this,

House Republicans recommended that the annual-cost-of-living adjustment for Social Security and other benefits tied to increases in the Consumer Price Index be reduced starting in 1999 [25]. Washington wisdom circulating at the time held that the CPI overstated inflation by as much as 1.5 percentage points and that a reduction in the CPI's value was not only justified, but defensible.

The problem was to give legitimacy to the Washington wisdom. Since the Bureau of Labor Statistics (BLS) moved cautiously and, from the point of view of the Washington establishment, unreliably in this matter, in June 1995 the Senate Finance Committee appointed a five-member panel of economists, chaired by former President Bush economic advisor Michael Boskin, to study the CPI and make recommendations on revisions. All members of what came to be called the Boskin Commission had respectable credentials and some might be described as eminent, but all had previously given Congressional testimony that the CPI exaggerated inflation. Economists who took a different view, such as former Commissioner of BLS Janet Norwood, were not invited to join the panel. In some quarters this is called stacking the deck.

The Boskin Commission released its report in early December 1996, claiming that consumer inflation was being overstated by the CPI by about 1.1% a year, arguing that the index did not adequately reflect the improving quality of goods, did not take into account new products quickly enough, did not properly reflect consumers' tendency to purchase cheaper alternatives when the price of goods rose, and did not properly take into account consumer shift toward discount stores [8].

A number of questions arise: What does it mean to say that the CPI overstates the reality of inflation by about 1.1% a year? Many interpret this to mean that there is an ideal standard for measuring the reality of inflation which is known by the Boskin Commission and that in comparing the BLS's CPI against this ideal standard, the BLS's CPI overstated inflation by about 1.1% a year.

> It is at this point that an understanding of the nature of math modeling is essential if the nonsense suggested by this interpretation is to be detected. **There is no ideal standard**.

The BLS's CPI is a valid conclusion of a math model based on data, accepted procedures, and **assumptions** made by the agency's economists. The Boskin Commission's proposed 1.1% per annum adjustment is based on the same data, same procedures, but with somewhat **different assumptions**. In effect, what the Boskin Commission was saying was that if you employ **our assumptions** rather than yours, then you have to make a 1.1% per annum downward adjustment in your CPI value. If your CPI increased by 3.3% over 1995, then 2.2% would be a more accurate description of the reality of inflation over that year, based on **our assumptions**.

Did the Boskin Commission consider the possibility that the BLS's CPI understates inflation? No; for discussion of this situation see [3, 15, 17, 24].

Is there good reason to prefer the **Boskin Commission's assumptions** over those made by the BLS? If politics were put aside, it becomes "experts" versus "experts." At a panel session at the annual meeting of the American Economic Association held in New Orleans in January 1997 Boskin and his four commission colleagues presented their views followed by BLS Commissioner Katherine Abraham who stated that 'she agreed with some of the Boskin Commission's recommendations, including that the CPI should be as close to a measure of the cost of living as possible; she added, however, that 'her agency would not and should not produce a CPI based partly on subjective judgments' [7]. Abraham later further elaborated to the Senate Finance Committee: 'If we get into the business of making judgments about things that are not measurable—guessing, even it's . . . a best guess—we really, I think, would be undermining the credibility of all the data we produce.'

The planting of an idea which developed into a BOLÉRO drum roll to reduce the CPI was probably done by the Congressional Budget Office when it asserted in late 1994 that the CPI exaggerated inflation by an amount between 0.2 and 0.8 percentage points a year. Federal Reserve Board head Alan Greenspan expressed the view that the CPI exaggerated inflation by an amount between 0.5 and 1.5 percentage points a year at a joint meeting of the House and Senate Budget Committees in January 1995. Greenspan also noted that correcting these estimates could save the government $150 billion over five years and suggested the possibility that Congress pass a law that would lower the CPI by a percentage point or half a percentage point for determining benefits tied to the CPI. It's a short hop from this plateau to the establishment of the Boskin Commission and what became the "official" view that the CPI overstated inflation by about 1.1 percentage points a year.

The Boskin Commission's report set BOLÉRO into motion in the form of a rash of calls to "fix" the way inflation is measured. Testifying before the Senate Finance Committee on January 30, 1997, Alan Greenspan recommended that an independent commission be established to set cost-of-living adjustments for federal receipts and outlays each year. Economist Martin Feldstein, who had been President Reagan's top economic advisor, suggested that Greenspan's proposed committee should recommend an "appropriate" inflation adjustment factor through informed judgment, apart from any adjustment made to the CPI by the Bureau of Labor Statistics through its normal work. Senators William Roth and Daniel Patrick Moynihan introduced a sense-of-the-Senate resolution that urged an accurate cost-of-living index. With momentum at a peak to push through a CPI fix, it all came apart. President Clinton, faced by strong opposition within his own party and constituencies like Labor and the elderly, decided not to pursue a CPI fix outside of the highly professional, non-political machinery of the Bureau of Labor Statistics. Republican enthusiasm for a "fix" waned with the discovery of a two-year old White House memorandum on how Democrats could use the issue against Republicans.

Within two years talk of engineering a CPI fix to help eliminate the budget deficit had turned to arguments over what to do with the projected budget surplus.

These were the salvos in a drive to strongly modify the math model which is the foundation stone on which the CPI rests.

10.9   Food for Thought Questions

1.   Suppose that in Tulsa a typical basket of food items cost $54 in 1996. At the same time the same basket of goods cost $61.50 in New Orleans, but only $50.25 in Omaha. Using the Tulsa 1996 rate as a base, construct an index that shows the relative overall food prices in New Orleans and Omaha at that time.

2.   Table 10.4 lists 2004 and 2007 prices in hundreds of dollars per year of some basic items needed to attend Ecap University.

Table 10.4

Item	Average Price	
	2004	2007
Tuition	75	96
Books	5	7
Transportation	3	4
Health Insurance	20	30

Construct the following and interpret your results:

(a)   An arithmetic mean of the price relatives, using 2004 = 100.

(b)   A simple aggregative index using 2004 as base year.

3.   The shellfish catch (in thousands of pounds) in Cape Crab during 2003 and 2008 is listed in Table 10.5

Table 10.5

Type of Shellfish	Price ($/lb.)		Quantity	
	2003	2008	2003	2008
Clams	0.23	0.56	26	43
Crabs	0.12	0.26	108	124
Lobsters	0.67	1.28	19	17
Oysters	0.51	0.62	29	27

Construct and interpret the following:

(a)   (i)   The 2008 simple aggregative price index.

       (ii)   The 2008 unweighted mean of quantity relatives index.

(b)   A weighted aggregative index comparing 2003 and 2008 prices, using 1998 quantities as weights.

(c)   A weighted arithmetic mean of price relatives, using current values as weights.

(d)   With 2003 = 100, a weighted mean of price relatives, using base year values as weights.

4.   Anna Lopez, accounting major at Hilevel University, pays her school costs by means of a part-time school job. Her wages were $80 per week in 2000 and $85 per week in 2001. However, college costs have increased 6% in 2001.

(a)   What is the price index for 2001 relative to 2000 = 100?

(b)   Calculate Anna's real weekly wage relative to 2000.

(c)   Does her 2001 salary give her more buying power, as compared to her 2000 salary? Explain.

5.   The consumer price index (CPI) and the average national hourly earnings in selected industries are listed in Table 10.6 for a sample of recent years.

Table 10.6

Year	CPI (2000 = 100)	Average Hourly Earnings		
		Service	Retail Trade	Manufacturing
2000	100.0	7.47	5.69	8.54
2002	107.3	8.16	6.03	8.92
2004	113.4	8.47	6.12	9.78
2006	120.0	8.81	6.27	10.77

(a)   Relative to 2000, determine the real hourly earnings of a typical worker in each of the three selected industries in 2004. What happened to the purchasing ability of each type of worker in 2004, relative to 2000? Explain.

(b)   Determine the equivalent 2002 hourly wage of a typical worker in each of the industries in 2006. In each case, does the 2006 wage earner have more or less purchasing power than he did in 2002? Explain.

(c)   Relative to 2004, determine the real hourly wage of a typical worker in each of the three fields in 2006. In each case, did average wages rise at a faster rate than did average prices during the 2004-2006 time span? Explain.

6.   The Gross Domestic Product (GDP) is often used as a measure of the economic activity of a country. The annual GDP of the United States (in $ billions) and the corresponding CPI for selected years is shown in Table 10.7.

Table 10.7

Year	GDP	CPI (1989 = 100)
1989	5438.7	100.0
1991	5916.7	109.8
1993	6553.0	116.5
1995	7253.8	122.9

(a)   Use the CPI to deflate the GDP for each of the years 1991, 1993, and 1995 to equivalent 1989 dollars.

(b)   Determine the percentage growth in real GDP during the time interval stated: (i) 1989 to 1993; (ii) 1991 to 1995; (iii) 1989 to 1995.

(c)   Compare the responses to part (b) with the corresponding growth rates of actual GDP during each of these time spans. What part may be attributable to inflation and what part expresses real growth in GDP?

7.   The average dividend in dollars per share of the Hirise stock fund for five consecutive years is shown in Table 10.8 along with an index of average dividend per share for competing stock funds of a similar kind, relative to 2004.

Table 10.8

	2004	2005	2006	2007	2008
Average Dividend Per Share	1.8	2.1	2.3	2.0	2.1
Index of Average Dividend Per Share (2004 = 100)	100	107	113	112	118

(a)   Compute the equivalent 2004 dividend per share for the Hirise stock fund in 2005. Was the increase in Hirise's 2005 dividend equal to the overall industry dividend increase, relative to 2004? Explain.

(b)   Determine the equivalent 2006 dividend per share for Hirise in 2007. Relative to 2006, was the decrease in Hirise's 2007 dividend in line with the overall industry average decrease? Explain.

(c)   Relative to 2004, did Hirise increase their dividend in 2007 to a larger extent than the overall industry average? Explain.

8.   The average costs of selected services for a sample of years in Busyville and the local Consumer Price Index for these years are listed in Table 10.9.

Table 10.9

Year	CPI (2000 = 100)	Accountant (hr.)	Lawyer (hr.)	Maid (hr.)
2000	100.0	40	80	10
2002	105.6	46	83	11
2004	115.3	51	95	12
2006	126.6	53	110	13

(a)   Relative to 2000, was the increase in the average legal fee rate at least as large as the overall rise in consumer costs in (i) 2002? (ii) 2004? Explain.

(b)   Determine the real average legal fee rate for 2002 and 2004, relative to 2000 costs.

(c)   Compared with 2000, was the increase in the average maid service rate equal to the overall rise in consumer costs in (i) 2002? (ii) 2006? Explain. Determine the real average maid service rate for each of these two years, relative to 2000.

(d)   Determine the real average accounting fee rate in 2006, relative to 2004. Compared with 2004, were 2006 average accounting fees in line with the average rise in consumer prices in 2006? Explain.

(e)   Determine the real average legal fee in 2006, compared to 2002. Were 2006 average legal fees in line with the average rise in prices in the 2004-2006 time interval? Explain.

9. When it is argued that the CPI exaggerated inflation by 1.5 percentage points (or by any figure for that matter), does this mean that there is an "ideal" standard against which the CPI is being compared and found wanting to the extent of 1.5 percentage points? Explain

10. Is it possible that the CPI underestimates inflation? Explain

10.10   References of Interest?

Concerning the CPI Brouhaha

1. W.J. Adams, "The CPI and You," *The Christian Science Monitor*, Dec. 10, 1996. This letter was in response to [11].

2. W.J. Adams, "Numbers Crunch," *The Economist*, Jan. 11, 1997. This letter is similar to [1].

3. D. Baker, "The Inflated Case Against the CPI," *The American Prospect*, Winter 1996.

4. D. Baker, "Hyping Hypo-inflation," *In These Times*, Jan. 6, 1997.

5. D. Baker, *Getting Prices Right: A Methodologically Consistent Consumer Price Index*, Economic Policy Institute, April 12, 1996.

6. J. Berry, "The CPI's Wide Reach," *The Washington Post National Weekly Edition*, December 23, 1996-Jan. 5, 1997.

7. J. Berry, "A Numbers Game Played for High Stakes: The BLS chief is standing fast," *The Washington Post National Weekly Edition*, March 17, 1997.

8. The Boskin Commission, *Toward a More Accurate Measure of the Cost of Living*, Final Report to the Senate Finance Committee from the Advisory Commission to Study the Consumer Price Index, Dec. 4. 1996.

9. M. Boskin, "More Accurate C.P.I," *The New York Times*, March 27, 1997, This letter is in response to [27].

10. J. Chait, "Revolution of '96", *The New Republic*, Dec. 30, 1996.

11. "An Enticing Number," *The Christian Science Monitor*, Dec. 10, 1996.

12. "Fix the CPI Yardstick Now," *The Christian Science Monitor*, March 7, 1997.

13. C. Duff, "Fix in Consumer Price Index Falls Short of Surgery Some Believe is Needed," *The Wall Street Journal*, April 1, 1996.

14. D. Francis, "Senators Eager to Fix Cost-of-Living Index," *The Christian Science Monitor*, Feb. 14, 1997.

15. D. Francis, "Fixing the Inflation Index—But Is It Really Broken?" *The Christian Science Monitor*, March 6, 1997.

16. D. Francis, "The Numbers Say Americans Made More . . ." *The Christian Science Monitor*, March 14, 1997.

17. D. Francis, "Poking Holes in the CPI Balloon," *The Christian Science Monitor*, March 14, 1997.

18. P. Grier, "Is US Economy Too Complex For Inflation Sleuths to Track?" *The Christian Science Monitor*, Dec. 5, 1996.

19. R. Hershey, Jr., "Labor Statistics Chief Takes Issue with Critics of Inflation Index," *The New York Times*, Dec. 20, 1996.

20. R. Kuttner, "Boskin's Magic Wand," *The Washington Post National Weekly Edition*, Dec. 16-22, 1996.

21. R. Kuttner, "The Fix is In on the C.P.I.," *The Washington Post National Weekly Edition*, March 17, 1997.

22. J. Madrick, "The Cost of Living: A New Myth," *The New York Review of Books*, March 6, 1997.

23. B. Moulton, "Bias in the Consumer Price Index: What is the Evidence?" *Journal of Economic Perspectives*, Fall 1996.

24. P. Passell, "Some Experts Say Inflation Is Understated," *The New York Times*, Nov. 6, 1997.

25. R. Pear, "G.O.P. Suggests Smaller Benefit Adjustments," *The New York Times*, May 11, 1995.

26. S. Pearlstein, "Fine-Tuning the Consumer Price Index," *The Washington Post National Weekly Edition*, Dec. 23, 1996-Jan 5, 1997.

27. J. Popkin, "Why Play a Numbers Game," *The New York Times*, Feb. 27, 1997.

28. R. Stevenson, "Moves Continue on Price Index Changes," *The New York Times*, March 5, 1997.

29. L. Uchitelle, "Balancing Quantity, Quality and Inflation," *The New York Times*, Dec. 18, 1996.

30. L. Uchitelle, "Measuring Inflation: Can't Do It, Can't Stop Trying," *The New York Times*, March 16, 1997.

31. L. Uchitelle, "The Negotiators Forgo a Cut in Inflation Index," *The New York Times*, May 3, 1997.

General References

1. R. Hershey, Jr., "How Good is Key U.S. Index," *The New York Times*, Sept. 28, 1984.

2. R. Samuelson, "How well does the Consumer Price Index measure the inflated prices we pay for dog food and doctors, parking lots and paperbacks?" *The New York Times Magazine*, Dec. 8, 1974.

3. P. J. McCarthy, "The Consumer Price Index", *Statistics: A Guide to the Unknown*, ed. J. M. Tanur *et al* (San Francisco: Holden Day, 1972, 266-275).

11

Probability Modeling

11.1   Preface

The focus of chapter 11 is on the concept of finite probability model and the issue of realism in its applications. Equally-likely-outcome models, Bernoulli trial models, and modeling centered on Bayes's theorem are particularly noteworthy. Subjective probabilities and the normal curves are considered from the probability modeling perspective.

11.2   The Concept of Finite Probability Model

Most books on finite mathematics, probability, and statistics introduce probability through the special case of equally-likely-outcomes followed by discussion (usually over-discussion, in my view) of counting tools to do business in this setting.

The problem with this approach I have found is that this initial exposure to probability comes to so dominate the mind-set of most students that regardless of what may be further said about probability for them probability is synonymous with counting.

Because of this my preference is to introduce the concept of finite probability model in neutral terms. I begin by introducing as background properties of relative frequencies of events connected with random processes and follow up by defining finite probability model as a structure born of these properties, but which goes beyond them. I find the analogy of parents and their child a helpful one to make. We often hear comments like, she has her mother's good looks but the temperament of her father, in recognition of similar characteristics that they share. But they are obviously distinct entities in their own right.

The following concept of finite probability model emerges from these considerations.

A **finite probability model** for a random process consists of two components:

(i)   A **sample space** $S = \{s_1, s_2, \ldots, s_n\}$.

(ii)   A function P, called a **probability function**, which assigns to each subset A of S a value, denoted by $P(A)$ and called the probability of A, subject to the conditions listed below. We state these conditions on the sample points and then extend them to subsets of S.

1. $P(s_1) \geq 0, \;\; P(s_2) \geq 0, \ldots, P(s_n) \geq 0$

2. $P(s_1) \leq 1, \;\; P(s_2) \leq 1, \ldots, P(s_n) \leq 1$

3. $P(s_1) + P(s_2) + \ldots + P(s_n) = 1$

4. If A is a subset of S, $P(A) =$ sum of the probabilities of the sample points describing A; if $A = \varnothing$, $P(A) = 0$. (For example, if $A = (s_1, s_5)$, $P(A) = P(s_1) + P(s_5)$.)

I further point out that the structural requirements of a probability model for a random process are somewhat analogous to a community's building code requirements for building our house. A building code does not tell us how to build our house. It tells us that however we build our house, for it to be legitimate in terms of the building code it must satisfy such and such conditions which are spelled out in the code. The concept of probability model does not tell us how to define a sample space and probability function for a random process; it tells us that in building a specific probability model for a random process, which can be done in many ways, we must satisfy the conditions stated in the definition of probability model (the building code in this case) in order for the model to be mathematically legitimate.

For details on this introduction to probability I recommend [2; ch. 6].

11.3   Which One is the *Right* Probability Model?

The follow-up of the afore observations is the issue of constructing specific probability models for specific random processes. To address this let us begin by returning to the man-in-the-park scenario presented in the Introduction.

One day the following conversation took place between P.M., a practical man, and S.P., a student of probability, who were sitting on a park bench.

    P.M. took a die from his pocket and said to S. P.: "Sir, I understand that you are a student of probability; would you help me with a problem?"

S.P.:   "What is your problem?"

P.M.:   "What is the probability of throwing an even numbered face with this die?"

S.P.:   "It's ½."

P.M.:   "How did you get ½?"

S.P.:   "Take the events 1 shows, 2 shows, 3 shows, 4 shows, 5 shows, 6 shows as your basic outcomes for the tossing of your die. There are six basic outcomes, three of which describe an even numbered face showing. Therefore, the probability that an even numbered face will show on a toss of your die is 3/6 which, of course, is ½."

P.M.:   "Are you sure about this?"

S.P.:   "No question about it; 3 divided by 6 is ½. The *precision* of math makes this conclusion *infallible*.

P.M.:   "I plan to participate in a 'friendly' game with this die tonight. What is the significance of this value for me?"

S.P.:   "If your die is tossed a large number of times, an even numbered face will show about 50% of the time; you can count on it."

P.M. participated in a 'friendly' game that evening and bet his money based on his expectation that an even numbered face would show about 50% of the time when his die is tossed a large number of times. His die was tossed 1000 times, but an even numbered face showed only 200 times, yielding the ratio 1/5, which deviates considerably from the predicted ratio close to ½. He found himself $400 poorer and, confused and angry, he went looking for S.P. where he had found him the day before. He found him on the same park bench and demanded an explanation of what had gone wrong and why math had failed him.*

Let us first address P.M.'s second question: "Why had math failed me?" which suggests the question, Did math fail him?, which takes us to P.M.'s assignment of probability $\frac{1}{6}$ to six events 1 shows, 2 shows, ... , 6 shows. P.M. is a victim of the equally-likely-outcome mind-set virus which compels him to assign the same probability value to all sample points in his sample space, no matter what. The equally-likely-outcome virus paralyzes thinking in much the same way that the polio virus paralyzes muscles.

An appropriate answer to P.M.'s question, "What is the probability of throwing an even numbered face with this die?" would be, how should I know, it's your die, not mine. However, if you are agreeable to the **assumption** that your die is well-balanced, of uniform construction, then the probability function that **best reflects this assumption** is the one that assigns the same value, $\frac{1}{6}$, to 1 shows, 2 shows, ... , 6 shows.

From this model we obtain the theorem that the probability an even numbered face shows is $\frac{1}{2}$.

---

*   Speaking of teachers whose influence is long lasting (see the quote from *The Education of Henry Adams*), I was introduced to the man-in-the-park scenario by my teacher Prof. Warren Hirsch many years ago in an introductory course on probability. It prompted a re-evaluation in my thinking about probability modeling.

As to "What had gone wrong?", it should be pointed out to P.M. that mathematics had not failed him. It had delivered what it is capable of—a theorem with respect to the equally-likely-outcome model, a theorem which proved (expensively) to be unrealistic.

There are two other scenarios that I have employed to introduce the subject of obtaining a specific probability model for a specific random process.

A Tale of Three Probability Models

Rasa Adams's friend Indre was on the eve of her twenty-first birthday for which a big family celebration was planned. Since Indre was interested in the laws of chance, her mother decided to obtain a gold die as a birthday present for her. On learning of this in a conversation with Indre's mother, Rasa concluded that the best present she could get Indre to go with the gold die would be a probability model; after all, what good is a gold die without an accompanying probability model? But what's the best place to shop for a probability model, thought Rasa. She finally decided to try Trump Models, run by the noted model builder Herman J. Trump. Rasa was not disappointed at Trump Models. There was a large selection with many attractive models, and Herman Trump was willing to custom design a model to your specifications if you so desired. Rasa finally narrowed her search to three models, $R4$ (red), $B7$ (blue) and $Y3$ (yellow). These models had the following specifications:

$$R4 \text{ (red): } S = \{1, 2, 3, 4, 5, 6\}; \ P(1) = \ldots = P(6) = 1/6$$

$$B7 \text{ (blue): } S = \{1, 2, 3, 4, 5, 6\}; \ P(1) = P(3) = P(5) = 1/9$$
$$P(2) = P(4) = P(6) = 2/9$$

$$Y3 \text{ (yellow): } S = \{1, 2, 3, 4, 5, 6\}; \ P(1) = P(3) = P(5) = 1/12$$
$$P(2) = P(4) = P(6) = 3/12$$

All three of these models are mathematically legitimate in that they all satisfy the probability model requirements. This is as far as Rasa can go in selecting a probability model for Indre's die. To determine which of these models, if any, is a realistic fit to Indre's die, she needed to know more about the nature of this die. Rasa called Indre's mother and asked this question. "I had to pay extra for this," said Indre's mother, "but this is an unusual die in that platinum weights have been inserted inside

it so that the even numbers are favored to show over the odd ones by 2 to 1." This led Rasa to purchase the blue model $B7$ as the most realistic model for Indre's die.

To examine a valid consequence of the blue model, consider the event E that an even number shows. $E = \{2, 4, 6\}$.

$$P(E) = P(2) + P(4) + P(6) = \frac{2}{9} + \frac{2}{9} + \frac{2}{9} = 0.67$$

The relative frequency interpretation of this result is that if a die whose behavior is described by the blue model is tossed a large number of times, an even number will show approximately 67% of the time.

After Indre's birthday it came to pass that her die was tossed 500 times. Records kept of the outcomes of the tossings reveal that an even number showed on 246 of the 500 tosses, so that the relative frequency of E is 0.492. Since the actual relative frequency of an even number showing, 0.492, is very much at variance with the projected value of approximately 0.67, this shows that a valid consequence of Indre's model, interpreted in relative frequency terms, is false. Indre's model is not a realistic one for her die, and her mother paid for an unusual die which was not delivered. The result observed suggests that the red model, based on the **assumption** that the die in question is well balanced, of uniform construction, is a more realistic one for the die that Indre actually obtained.

The Adventures of Hasty Harry

What should I get my brother for his birthday, pondered Bottom-line Bob. It's his thirtieth, the big 30, and this calls for something very special. His brother Harry, who was fond of games of chance, and had an extensive collection of "unusual" dice. Bob decided to add to it by obtaining for him what would undoubtedly be the crown jewel of his collection, a gold die embedded with diamond chips to show off the spots on its faces. Bob had the die custom made and on the appointed day a very pleased Harry received a very special die.

Harry could hardly wait to show off his new treasure to his friends and, in addition, win some vacation money in a bit of friendly gaming activity that was certain to follow. In preparation for this he went to Martin's Models to obtain a probability model to describe the behavior of the new crown jewel of his collection.

"Mr. Martin, I want you to build me a probability model for the tossing of this die. I'm particularly interested in the probability that an even number shows."

"What can you tell me about your die, Harry? What do you know about its behavior from your experience with it?"

"I just got it as a present and I have no experience with it. I want to be prepared with a probability model before obtaining that experience. A die is a die, nothing special, apart from its being made of gold with diamond chips. What else is there to know?"

"All right Harry, I'll proceed on the **assumption** that it's an ordinary die of uniform composition, a fair die, as we say. This being the case the equally likely outcome model R4 (red) that I have in stock is the most realistic description of the behavior of your die.

R4 (red): $S = \{1, 2, 3, 4, 5, 6\}$

$$P(1) = P(2) = \ldots = P(6) = \frac{1}{6}$$

It follows from R4 that the probability that an even number shows is:

$$P(2) + P(4) + P(6) = \frac{1}{6} + \frac{1}{6} + \frac{1}{6}$$
$$= \frac{3}{6}$$
$$= 0.50$$

"What does this mean in terms of some friendly gaming activity, Mr. Martin?"

"If you toss this die a large number of times, an even number should show roughly half the time. We cannot say when an even number will show, that's a matter of chance, but it should show roughly 50% of the time."

Reality Strikes

Harry proudly showed his die to his friends and all, with the exception of Harry, had a good time playing games of chance in which "friendly" bets were placed on which face would show when the die was tossed. Harry expected an even number to show roughly 50% of the time as predicted by Martin's model and bet accordingly. It came to pass, however, that after 500 tosses of his die an even number had showed 66% of the time, which is sharply at variance with what Harry had expected. He had hoped to make a modest profit from this "friendly" gaming activity and now he found himself an unfriendly three hundred bucks in the red. Confused and feeling that he had been cheated, he stormed back to Martin's models for some answers.

"I don't understand what went wrong Martin. If your mathematics is so precise, how could it happen that an even number showed 66% of the time instead of around the 50% you told me to expect? I'm three hundred bucks down because of this. You sold me a defective model and I want my money back."

"Mathematics, the probability model I gave you in this case, Harry, delivered what it was capable of, namely, a valid conclusion with respect to the **assumptions** made. Please remember that it was you who provided me with the starting point of the analysis. I quote: 'a die is a die, nothing special.' As it turned out there was something very special about this die which **made my assumption, based on your information, unrealistic**. As a result we obtained a valid conclusion from the model which, when interpreted in relative frequency terms, tells us about how often an even number can be expected to show. This proved to be at variance with the nature of your die.

> When there is a sharp conflict between a math model and the reality it is intended to describe, reality always wins.

Didn't your brother say anything to you about the nature of the die, or were you too dazzled by the gold and diamond chips to pay attention?"

"I'm not sure now. I'll have to ask him. Maybe I was too hasty."

A New Model for Hasty Harry's Die

"Mr. Martin, I spoke to my brother and I listened this time. He said he told me that he had the die weighted internally so that the even numbered faces were twice as likely to show as the odd numbered ones. It's not at all an ordinary die in terms of its internal make up."

"I spoke to Bob after you left and he told me about the die's structure. He spent a lot of money to have the die made in this way and, ironically, it ended up costing you money. I developed another probability model for your die based on the information Bob gave me. This model should be a much better fit to your die.

I call it B7 (blue); it is defined by,

$$B7 \text{ (blue)}, S = \{1, 2, 3, 4, 5, 6\}$$
$$P(1) = P(3) = P(5) = \frac{1}{9}$$
$$P(2) = P(4) = P(6) = \frac{2}{9}$$

It follows as a valid conclusion (theorem) from B7 that the probability an even number shows is:

$$P(2) + P(4) + P(6) = \frac{2}{9} + \frac{2}{9} + \frac{2}{9}$$
$$= \frac{2}{3} = 0.67$$

The relative frequency interpretation of this valid conclusion is that if your die were tossed a large number of times, an even number would show about 67% of the time. This, I understand, is in agreement with the "friendly" game experience that you had."

"How much do I owe you?"

"The same as for the previous model. There's no additional charge for your hasty action. You've paid that price."

Hasty Harry's Other Die

"I have another die in my collection Mr. Martin, one that is weighted in such a way that the even numbered faces are three times as likely to show as the odd numbered ones. Do you have a suitable model for this die?" "As a matter of fact I do. It's Y3 (yellow) in my catalog listing and it's defined by:

$$Y3 \text{ (yellow)}, \quad S = \{1, 2, 3, 4, 5, 6\}$$

$$P(1) = P(3) = P(5) = \frac{1}{12}$$
$$P(2) = P(4) = P(6) = \frac{3}{12}$$

A Food for Thought Question

1. Let E denote the event that an even number shows.

    (a)   Find $P(E)$.

    (b)   State the relative frequency interpretation of the result obtained in (a).

    (c)   In tossing the other die in Harry's collection 1000 times, an even number was observed to show 665 times. Does this show that the conclusion obtained in (a) is not valid? Explain.

    (d)   Is the conclusion reached in (a), interpreted in relative frequency terms, true? Explain.

    (e)   Is the yellow model Y3 realistic for the afore die? Explain.

Susan's Problem

Susan's problem, as I term it, provides us with a setting to explore different approaches to formulating a probability model for a random process and addressing the question of what is *right* and *wrong* about them.

Susan Reti was interested in determining the probability that an even sum shows for the process of tossing a pair of well-balanced dice of uniform construction (one red and one green). She asked two of her friends, Rachael and Laura, if they would help her set up probability models to determine the probability of this event. Both were glad to do so.

Rachael's Model: Rachael took as her sample space S_1 the set of events described by all ordered pairs of integers between 1 and 6, inclusive, as shown in Table 11.1. The first number in each ordered pair specifies the number which shows on the red die and the second number specifies the number which shows on the green die.

Table 11.1

$$S_1 = \begin{cases} (1,1) \ (1,2) \ (1,3) \ (1,4) \ (1,5) \ (1,6) \\ (2,1) \ (2,2) \ (2,3) \ (2,4) \ (2,5) \ (2,6) \\ (3,1) \ (3,2) \ (3,3) \ (3,4) \ (3,5) \ (3,6) \\ (4,1) \ (4,2) \ (4,3) \ (4,4) \ (4,5) \ (4,6) \\ (5,1) \ (5,2) \ (5,3) \ (5,4) \ (5,5) \ (5,6) \\ (6,1) \ (6,2) \ (6,3) \ (6,4) \ (6,5) \ (6,6) \end{cases}$$

The information that the dice are well-balanced is best reflected by the probability function P which assigns the same value, 1/36, to each sample point in S_1.

The sample points with even sums are found in alternate diagnols of Table 11.1, beginning with the first, and are 18 in number. Thus, from Rachael's model it follows that the probability of E, that an even sum shows, is:

$$P(E) = \frac{18}{36} = 0.50$$

The relative frequency interpretation of this conclusion is that if a pair of well-balanced dice are tossed a large number of times, an even sum will show approximately 50% of the time.

Laura's Model: Laura proceeded to analyze Susan's problem in a different way. She took as her sample space

$$S_2 = \{2, 3, 4, 5, 6, 7, 8, 9, 10, 11, 12\},$$

where 2 is the event that the sum of the numbers showing is 2, 3 is the event that the sum of the numbers showing is 3, and so on.

The **assumption** that the dice are well-balanced led Laura to the probability function P which assigns the same value, 1/11, to the eleven sample points in S_2.

$$E = \{2, 4, 6, 8, 10, 12\}, \text{ and Laura obtained}$$

$$P(E) = \frac{6}{11} = 0.55$$

for the probability that an even sum shows.

The relative frequency interpretation of this result is that if a pair of well-balanced dice are tossed a large number of times, and even sum will show approximately 55% of the time.

Susan was more confused than ever and turned to her cousin Jack for help with her problem.

Jack's Model: Jack took as his sample space S_3 the collection of events shown in Table 11.2.

Table 11.2

{1,1}, {1,2}, {1,3}, {1,4}, {1,5}, {1,6}
{2,2}, {2,3}, {2,4}, {2,5}, {2,6}
{3,3}, {3,4}, {3,5}, {3,6}
{4,4}, {4,5}, {4,6}
{5,5}, {5,6}
{6,6}

Here {1,1} is the event that both dice show 1, {1,2} is the event that one die shows 1 and the other shows 2, and so on. There are 21 sample points, Jack observed. Since the dice are **assumed** to be well-balanced, Jack was led to take as his probability function P the one which assigns the same value, 1/21, to each sample point. There are 12 sample points with even sum and this led Jack to conclude that $P(E)$, the probability that an even sum shows, is 12/21 or 0.57.

The relative frequency interpretation of this result is that if a pair of well-balanced dice are tossed a large number of times, an even sum will show approximately 57% of the time.

P (the sum showing is even) = 0.50 or is it 0.55 or is it 0.57, or none of the preceding, Susan asks, of course we would have to reply that all three probability values are *right* in the sense of being **valid with respect to their respective probability models**, but not all three, perhaps none, are *right* in the sense of being realistic.

Tossing a pair of dice, certified to be well-balanced, a large number of times and recording the outcome establishes that Rachael's model is *right* in the sense of being realistic while the others are *wrong* in this sense.

The lesson of this and the afore examples is that as appealing as an equally-likely-outcome model might be in terms of (relative) simplicity it does not necessarily serve us well in terms of realism.

Food for Thought Questions

2. Consider the model $G4$ (green) for the process of tossing a certain die. $S = \{1, 2, 3, 4, 5, 6\}$,

$$P(1) = P(3) = P(5) = \frac{1}{10}, \quad P(4) = \frac{3}{10}, \quad P(2) = P(6) = \frac{2}{10}.$$

Let A denote the event that an odd number shows.

(a)   Find $P(A)$.

(b)   State the relative frequency interpretation of the result obtained in (a).

(c)   In tossing the die in question 1000 times an odd number was observed to show 302 times. Does this evidence establish that the conclusion obtained in (a) is valid? Explain.

(d)   Is the conclusion obtained in (a), interpreted in relative frequency terms, true? Explain.

(e)   Is the green model $G4$ realistic for the die in question? Explain.

3. The appearance of a valid conclusion that is false alerts us to the unrealistic nature of **Jack's assumption** of equally-likely outcomes for S_3.

(a)   Pinpoint the difficulty with Jack's assumption.

(b)   How should Jack's probability function be modified to make it realistic for well-balanced dice?

4. How should Laura's probability function be modified to make it realistic for well-balanced dice?

5. The Twolow Company makes light bulbs. Two plants, $P1$ and $P2$, carry out the production process. The daily output is 8000 bulbs, with $P1$ producing

5000 bulbs of which 1 percent are defective, and *P2* producing 3000 bulbs of which 0.5% are defective. A bulb is selected from the day's output.

(a)   When asked to determine the probability that a defective bulb is chosen, Mark Twolow set up the following probability model for the selection process, based on the **assumption** that the bulb is selected at random from the day's output. $S = \{GP1, GP2, DP1, DP2\}$, where *GP1* is the event that a good bulb made by *P1* is selected, and so on.

$$P(GP1) = P(GP2 = P(DP1) = P(DP2) = \frac{1}{4}$$

Mark determined the probability that a defective bulb is chosen to be $P(DP1) + P(DP2) = 1/2$. (1) Is Mark's model satisfactory? How so? (2) Is Mark's conclusion *correct*? How so?

(b)   Mark's brother Bob suggested a simpler model, again based on the **assumption** that the bulb is chosen at random from the day's output. $S = \{G,D\}$, where *G* is the event that a good bulb is selected and *D* is the event that a defective bulb is chosen. $P(G) = P(D) = 1/2$. Bob pointed out that he and Mark had reached the same conclusion, so that each confirms the *correctness* of the other. (1) Would you agree or disagree with Bob's comment? Explain. (2) Is Bob's model satisfactory? Explain. (3) Is Bob's conclusion correct? (4) When asked for the probability that a defective bulb made in *P1* is selected in terms of his model, Bob gave 1/4 as the answer. Would you agree or disagree? How so?

(c)   Should Mark's probability function be modified? How so? If modification is called for, how would you carry it out?

(d)   Should Bob's probability function be modified? Explain. If modification is called for, how would you carry it out?

6. Jason took a well-balanced coin from his pocket and asked his friend Andrew to help him determine the probability of throwing one head and one tail on two successive tosses of the coin. Andrew took $S = \{0,1,2\}$, where 0 is the event that no heads show in the two tosses, 1 is the event that one head shows in the two tosses, etc., as his sample space. He

defined a probability function P by $P(0) = P(1) = P(2) = 1/3$, so that the probability of throwing one head and one tail is $1/3$.

The relative frequency interpretation of this conclusion is that if a well-balanced coin is tossed twice in succession a large number of times, a head and tail will show approximately 33.3% of the time. When Jason's well-balanced coin was tossed twice in succession 500 times (which involves 1000 tosses), a head and tail were observed to show 246 times.

(a) Does this mean that Andrew's conclusion is not valid? How so? (Red Herring Alert)

(b) Is Andrew's conclusion, interpreted in relative frequency terms, true? How so?

(c) How is the discrepancy between the predicted relative frequency and the observed relative frequency to be explained?

(d) Set up your own probability model for tossing a coin, assumed to be well-balanced, twice in succession.

(e) Should Andrew's probability function be modified? Explain. If modification is called for, how would you carry it out?

7. Bill Albert succeeded in finding a rare pair of loaded dice. With reference to the set S_1 of 36 outcomes described in Table 11.1, these dice have the property that outcomes with an even sum [such as (1,1), (1,3), and so on] are twice as likely to occur as outcomes with an odd sum [such as (2,1), (3,2), and so on].

(a) Define a probability function on S_1 that best reflects the nature of Bill's dice.

(b) Find the probability that (i) an even sum shows; (ii) a sum of 7 shows; (iii) a sum less than 6 shows.

8. Jack Jones was interested in the number of times head shows in three successive tosses of a well-balanced coin. He took as his sample space $S =$ {0, 1, 2, 3}, where 0 is the event that no head shows in the three tosses, 1 is the event that one head shows in the three tosses, and so on. Jack assigned equal probabilities of ¼ to these sample points and determined the probability of head showing twice in the three tosses to be ¼.

    The relative-frequency interpretation of this conclusion is that, if the coin is tossed three times in succession a large number of times, head will show exactly twice approximately 25 percent of the time. Yet when Jack tossed a well-balanced coin three times in succession a large number of times, he observed that head showed exactly twice 38 percent of the time. In light of these developments, consider the following questions.

    (a) Isn't Jack's conclusion valid? Explain. (Red Herring Alert)

    (b) Isn't Jack's conclusion true? Explain.

    (c) How is the discrepancy between the predicted relative frequency and the observed relative frequency to be explained?

9. For his birthday, Herman received a pair of dice that, he was told, were well-balanced. On the basis of the probability model that assigns the same value, $\frac{1}{36}$, to each of the sample points in the usual 36-element sample space for the process, Herman determined the probability of an even sum showing to be 0.50. He expected that an even sum would show in the neighborhood of 500 times for 1000 tosses of his dice and made betting plans accordingly. Herman participated in a friendly game one evening, and after 1000 tosses of his dice an even sum had showed 200 times and Herman was $600 poorer. Disappointed, confused, and angry, Herman raised the following questions.

    (a) If mathematics is such a *precise* subject, how could this happen?

    (b) Isn't my conclusion *correct*?

    (c) What went *wrong*?

11.4   Equally-Likely Outcome Modeling and Syndrome

Let us return to the man-in-the-park scenario:

One day the following conversation took place between P.M., a practical man, and S.P., a student of probability, who were sitting on a park bench.

P.M. took a die from his pocket and said to S. P.: "Sir, I understand that you are a student of probability; would you help me with a problem?"

S.P.:   "What is your problem?"

P.M.:   "What is the probability of throwing an even numbered face with this die."

S.P.:   "It's ½."

P.M.:   "How did you get ½?"

S.P.:   "Take the events 1 shows, 2 shows, 3 shows, 4 shows, 5 shows, 6 shows as your basic outcomes for the tossing of your die. There are six basic outcomes, three of which describe an even numbered face showing. Therefore, the probability that an even numbered face will show on a toss of your die is 3/6 which, of course, is ½."

P.M.:   "Are you sure about this?"

S.P.:   "No question about it; 3 divided by 6 is ½. The *precision* of math makes this conclusion *infallible*.

S.P. suffers from **equally-likely outcome syndrome**—compulsive behavior that leads one to take the ratio of the number of sample points describing an event to the total number of sample points as the probability of the event, no ifs-ands-or-buts.

It is an infirmity that, in my experience, afflicts many students and teachers of mathematics, as well. I regard as my greatest challenge in teaching probability to break the hold that the equally-likely outcome syndrome has on the thinking of so many of my students.

The "loaded" die is a familiar setting that can be employed to initiate the breaking down of the equally-likely outcome syndrome. But, like the Berlin Wall, it does not yield easily and many sledge hammers are needed.

The only feasible road to success is through example, example, example followed by question for thought, question for thought. One scenario that I employ is the one mile race.

The One Mile Race

The Alumni Association of Ecap College has organized a one mile race to be run by two faculty, W.J. Adams and H. Lurier, and two alumni, J. Ross and E. Kapp. What is the probability that Adams finishes first?

One approach to this problem is to note that there are 4 x 3 x 2 x 1 = 24 possible finishes, that Adams is first in 3 x 2 x 1 = 6 of them, and conclude that the probability Adams finishes first is 6/24 = 0.25.

This approach, based on the underlying **assumption of equally-likely-outcomes**, reduces to counting—count the number of ways Adams can finish first, the total number of possible finishes, and divide.

> If we dig a bit deeper we find that the **assumption of equally-likely outcomes** is based on the **assumption** that the four participants in the race are in comparable physical condition, have comparable racing experience, and that the conditions under which the race is to be run (outdoor temperature, for example) do not favor some participants over others.
>
> When we learn that Adams is out of shape and tires easily, it becomes clear that the **equally-likely outcome assumption** is untenable.

Is Dealing a Card Necessarilly as Straightforward as it Sounds?

Consider the process of dealing a card from a standard deck of 52 cards and let us address the problem of setting up a probability model for this process and determining the probability that a picture card is dealt.

For convenience in referring to the cards, let us set up a translation system so we can refer to the cards as 1,2, ... , 52; in this translation system, 1 might denote the ace of spades, 2 king of spades, etc. We take as our sample space,

$$S = \{C_1, C_2, \ldots, C_{52}\},$$

where C_1 is the event that card 1 is dealt, C_2 is the event that card 2 is dealt, etc.

Our next task is to define a probability function P on S. This can be done in many ways, and the function P that emerges depends on the assumption we make about how the card will be dealt from the deck.

Suppose we **assume** what is usually assumed in such a situation, but not always made explicit, that the card is dealt from a well-shuffled deck in an unbiased way, at random, as we say. The probability function P which best reflects this **assumption** assigns the same value, 1/52, to each sample point in S.

We thus emerge with the following probability model:

Model 1 : $S = \{C_1, C_2, \ldots, C_{52}\}$

$$P(C_1) = \cdots = P(C_{52}) = \frac{1}{52}$$

From Model 1 it follows that the probability that a picture card is dealt is

$$\frac{12}{52} = 0.23$$

As is well-known, some card dealers are less than honest. Suppose we **assume**, based on past experience, that the dealer intends "to arrange things" so that the card we are dealt is neither a picture card nor an ace, but that any of the other cards may be dealt without bias. For notational convenience suppose that the picture cards and aces in the deck are numbered 1,2, . . . 16, and that the cards 17,18 . . . . 52 correspond to the remaining cards. This leads to the Model 2.

Model 2: $S = \{C_1, \ldots C_{16}, C_{17}, \ldots C_{52}\}$,

$$P(C_1) = \cdots = P(C_{16}) = 0, \; P(C_{17}) = \cdots = P(C_{52}) = \frac{1}{36}$$

From Model 2 it follows that the probability that a picture card is dealt is 0.

Which model is "correct"? Correct, I emphasize to my students, is a word that we must always consider with caution because of its double edged nature.

In the sense of satisfying the conditions that define a probability model, both are correct. The conclusions obtained from these models are correct in the sense of being **valid** with respect to each of them.

Which model, if either, is correct in the sense of being **realistic**? is another question I further emphasize. Only real-world experience with cards can help us address this question.

For those who have been brought up on equally-likely outcomes this is a very hard sell.

The Poker Hand

A poker hand of 5 cards is to be dealt from a standard deck of 52 cards. What is the probability that the hand contains 3 aces?

For those having been drilled in counting techniques the response is

$$\frac{C(4,3)\cdot C(48,2)}{C(52,5)} = 0.0017,$$

which, of course, is *correct*, strings attached. It's the strings attached that I insist my students give me. I expect a probability model together with an **assumption** that backs it, such as in the following:

Model 1

Envision, for simplicity and convenience, the cards tagged $C_1, C_2, \ldots, C_{52}$. Take as your sample space,

$$S = \{(C_1, C_2, C_3, C_4, C_5), \ldots, (C_{48}, C_{49}, C_{50}, C_{51}, C_{52})\}$$

based on all combinations of 5 cards that can be chosen from 52, numbering $C(52, 5)$.

Based on the **assumption** that the hand is dealt at random—no bias, intended or inadvertent, that favors some hands being dealt over others—we take as our probability function P the one that assigns the same value, $\frac{1}{C(52,5)} = \frac{1}{2,598,960}$, to all sample points in S.

From this **string attached** it follows that:

$$P(3 \text{ aces}) = \frac{C(4,3)\cdot C(48,2)}{C(52,5)} = 0.0017$$

I insist on this **string attached** being described in answer to the question, what is the probability that the hand dealt contains 3 aces? If on an exam the only response is

$$\frac{C(4,3) \cdot C(48,2)}{C(52,5)},$$

out of 10 points I would give it, perhaps 5.

I further hammer away at the equally-likely outcome syndrome by asking my students to consider the following situation:

Let us suppose that it had been "arranged" with the dealer that he deal the poker player a hand consisting of three aces.
   Clearly Model 1 is **not realistic** for this situation.

Model 2

Assumption

The three aces are dealt at random from the four and the two non-aces are dealt at random from the remaining non-aces.

In formulating Model 2 we shall stay with the same sample space as the one employed for Model 1, but modify the probability function.
   There are $C(4,3) = 4$ ways of choosing 3 of 4 aces and $C(48,2) = 1128$ ways of choosing 2 of 48 non-aces. Therefore there are $C(4,3) \cdot C(48,2) = 4512$ ways of choosing 3 of 4 aces and 2 of 48 non-aces.
   The modified probability function, call it P_M, which realistically reflects the afore circumstance and **assumption** assigns 0 to the hands not consisting of 3 aces and 2 non-aces and

$$\frac{1}{C(4,3) \cdot C(48,2)} = \frac{1}{4512}$$

to the hands consisting of 3 aces and 2 non-aces.

The probability that the hand dealt contains 3 aces is, of course 1, which is the sum of the probabilities of the 4512 hands containing 3 aces.

   One of my colleagues, an excellent teacher in many ways, was brought up on combinatories as the exclusive focus in his study of probability, as was I and almost all students, past and present, who studied/study probability. He finds combinatonics irresistible because of its challenging problem solving dimension and he enjoys sharing this dimension with his students.

The price of this approach to teaching probability with the main focus on combinatories is high. The math modeling perspective is submerged, with chilling consequences. I have found that if you ask a student who had been indoctrinated by combinatorics ideology in high school, how many hands of 5 cards dealt from a standard deck of 52 cards contain 3 aces, in too many cases you will not be given C(4,3).C(48,2) as the answer, but

$$\frac{C(4,3) \cdot C(48,2)}{C(52,5)},$$ the probability that 3 aces are dealt under, of course, suitable strings attached.

And then there are those who come to believe that the probability of 3 aces being dealt is:

$$\frac{C(4,3).C(48,2)}{C(52,5)}$$

period. Model 2 in the afore poker hand problem and questions 10 and 11 type food-for-thought questions are intended to address such misunderstandings.

Population Estimation Problems

> How many fish are in your favorite lake? How many raccoons are in your neighborhood? How many animals of your favorite kind are in the game reserve or national park? More generally, how many "whatever" are in your region of interest?

These problems bring together combinatorics and the urgent need to carefully give thought to the **realism of the assumptions** made.

One approach to problems of this sort is based on what is called the **capture-release-recapture method.** We illustrate it by considering a fish population estimation problem, but the approach is applicable to the other situations noted as well.

We begin by catching a certain number of fish from the lake—100, say. These fish are tagged so as to be identifiable if caught again and are thrown back into the lake. We wait for a reasonable time to elapse to allow the fish to disperse (maybe a few days, and then catch another batch of fish, 200, say, and make note how many in this batch were caught before. Let us say that one fish was twice caught.

Let N denote the number of fish in the lake. Our problem is to estimate N. To do this we set up a probability model for the process of catching the second batch of 200 fish. As our sample space we take the event represented by the collection of all batches (combinations) of 200 fish that can be selected from N. There are $C(N,200)$ such batches.

> **Based on the assumption that all fish in the lake have the same likelihood of being caught,** we take as our probability function P the one that is the same value,
>
> $$\frac{1}{C(N,200)}$$
>
> to each sample point. We next determine the probability of catching 1 marked fish.

This probability is equal to:

$$\frac{\text{number of ways of catching 1 marked fish}}{C(N,200)}$$

The number of ways of catching 1 marked fish in a batch of 200 fish is equal to the number of ways of catching 1 of 100 marked fish. $C(100,1)$, times the number of ways of catching 199 of N - 100 unmarked fish, $C(N-100, 199)$, which yields the product $C(100, 1).C(N - 100,199)$. Thus:

$$P(1 \text{ marked fish is caught}) = \frac{C(100, 1).C(N - 100, 199)}{C(N, 200)} \qquad (1)$$

The right side of (1) depends on N. It varies as different numbers are substituted for N. Of particular interest is that number which when substituted for N makes (1) assume its maximum level. This value is called the **maximum likelihood estimate of N;** that is, the maximum likelihood estimate of N is that number which maximizes the probability of catching the number of marked fish that were actually caught in the second batch.

We shall now show that the maximum likelihood estimate of N is 20,000.

The right side of (1) is a function of N, which we shall denote by $P(N)$.

$$P(N) = \frac{C(100,1) \cdot C(N-100,199)}{C(N,100)} \qquad (2)$$

We seek a positive integer value of N such that.

$$P(N-1) \leq P(N) \quad and \quad P(N) \geq P(N+1) \qquad (3)$$

or, equivalently:

$$\frac{P(N-1)}{P(N)} \leq 1 \quad and \quad \frac{P(N)}{P(N-1)} \geq 1 \qquad (3)$$

Our first task is to determine and simplify P(N - 1), P(N), and P(N + 1). From (2) we obtain

$$P(N) = \frac{\dfrac{100(N-100)...(N-298)}{199 \cdot 198...1}}{\dfrac{N(N-1)...(N-199)}{200 \cdot 199...1}}$$

$$= \frac{100(N-100)...(N-298.}{199.198...1} \; \frac{200.199.1}{N(N-1)...N-199)}$$

$$= \frac{100(200)(N-100)...(N-298)}{N(N-1)...(N-199)} \qquad (4)$$

For P(N - 1) we have:

$$P(N-1) = \frac{C(100,1) \cdot C(N-1-100, \ 199}{C(N-1, \ 200)}$$

$$= \frac{\dfrac{100(N-101)...(N-299)}{199.198...1}}{\dfrac{(N-1)...(N-200)}{200.199...1}}$$

Inverting and simplifying yields:

$$P(N-1) = \frac{100(200)(N-101)...(N-299)}{(N-1)...(N-200)} \tag{5}$$

For P(N + 1) we have:

$$P(N+1) = \frac{C(100,1) \cdot C(N+1-100, \ 199)}{C(N+1, \ 200)}$$

$$= \frac{\dfrac{100(N-99)...(N-297)}{199 \cdot 198...1}}{\dfrac{(N+1)...(N-198)}{200 \cdot 199...1}}$$

Inverting and simplifying yields:

$$P(N+1) = \frac{100(200)(N-99)...(N-297)}{(N+1)...N-198)} \tag{6}$$

From (4) and (5) we have:

$$\frac{P(N-1)}{P(N)} = \frac{\dfrac{100(200)(N-101)...(N-299)}{(N-1)...(N-200)}}{\dfrac{100(200)(N-100)...(N-298)}{(N-1)...(N-199)}}$$

Inverting and canceling like terms yields:

$$\frac{P(N-1)}{P(N)} = \frac{100(200)(N-101)...(N-299)}{(N-1)...(N-200)} \cdot \frac{N(N-1)...(N-199)}{100(200)(N-100)...(N-298)}$$

$$\frac{P(N-1)}{P(N)} = \frac{(N-299)N}{(N-200)(N-100)} \tag{7}$$

From (4) and (6) we have:

$$\frac{P(N)}{P(N+1)} = \frac{\dfrac{100(200)(N-100)...(N-298)}{N(N-1)...(N-199)}}{\dfrac{100(200)(N-99)...(N-297)}{(N+1)...(N-198)}}$$

Inverting and canceling like terms yields:

$$\frac{P(N)}{P(N+1)} = \frac{100(200)(N-100)...(N-298)}{N(N-1)...(N-199)} \cdot \frac{(N+1)...(N-198)}{100(200)(N-99)...(N-297)}$$

$$\frac{P(N)}{P(N+1)} = \frac{(N-298)(N+1)}{(N-199)(N-99)} \tag{8}$$

From (3), (7), and (8), our problem reduces to finding N such that:

$$\frac{(N-299)N}{(N-200)(N-100)} \leq 1 \ and \ \frac{(N-298)(N+1)}{(N-199)(N-99)} \geq 1$$

From the first of this conditions we obtain:

$$(N-299)N \leq (N-200)(N-100)$$

$$N^2 - 299N \leq (N^2 - 300N + 20{,}000)$$

$$N \leq 20{,}000$$

From the second of these conditions we obtain:

$$(N-298)(N+1) \geq (N-199)(N-99)$$

$$N^2 - 297N - 298 \geq N^2 - 298N + 19{,}701$$

$$N \geq 19{,}999$$

Thus:

$$\boxed{19{,}999 \leq N \leq 20{,}000}$$

We may take 19,999 or 20,000 as our maximum-likelihood estimate of the size of the fish population.

More generally, the maximum—likelihood function that corresponds to catching k fish from the lake, tagging them and throwing them back, and catching a second batch of n fish that is observed to contain r tagged fish is given by:

$$P(N) = \frac{C(k,r) \cdot C(N-k, n-r)}{C(N,n)}$$

In our example k = 10, n = 200, and r = 1.

By an analysis similar to the preceding one, it can be shown that the maximum-likelihood estimate of N is characterized by:

$$\frac{nk}{r} - 1 \leq N \leq \frac{nk}{r}$$

Black Bear Sightings

There have been more black bear sightings around the town of Charlotte and its residents have come to raise questions about the size of this population in the surrounding forest.

At the request of town's council an applied statistic team at the local college undertook the project of obtaining a maximum likelihood estimate of the bear population. Six bears were caught, tagged and released. Shortly thereafter 5 bears were caught and 1 had been previously caught.

In the situation k = 6, n = 5, and r = 1. thus, the maximum-;likelihood estimate of the bear population is:

$$N = \frac{nk}{r} = \frac{5(6)}{1} = 30$$

Beware the assumption

As *precise* as the maximum likelihood estimate of an animal population might impress us as being, we must not allow ourselves to forget that what has been established is **its** *precision* **in a mathematical sense** (valid conclusion).

Its *precision* **in a reality sense** hinges on the realism of the second sample of animals caught being a random sample of the population. The **realism of this assumption** depends on how well dispersed the tagged animals are throughout the population. **If this dimension is open to question, so is the realism of the maximum likelihood estimate of the population's size.**

This is the crucial point that I emphasize to my classes, one that takes us to the problem of achieving random sampling in practice as envisioned in theory. For discussion of this issue see sec 20.3.

Food-for-thought questions 13 and 14 offer opportunities to grapple with questions arising from the afore.

Food for Thought Questions

10. Arnold Wilson, newly appointed Secretary of the recently established Department of Mathematical Affairs, was called on to make a few remarks at his installation as Department Secretary.

    "One thing that separates mathematics from politics is that mathematics is definite, no ten sides to every story. Consider, for example, the event that there are 3 kings in a hand of 5 cards dealt from a standard deck of 52 cards. Its probability is

    $$\frac{C(4,3) \cdot C(48,2)}{C(52,5)},$$

    and that's a fact, no debate. If I submitted this statement in the form of a resolution before any legislative body it would pass unanimously."

    If you were a member of a legislative body to which the afore statement was submitted as a resolution, would you vote for it? Explain.

11. Set up a probability model for the process of dealing a poker hand of 7 cards from a standard deck of 52 cards. State the **assumption** that underlies your model.

    (a) Find the probability that the hand dealt contains 2 aces.

    (b) Suppose that it had been "arranged" with the dealer that the hand to be dealt was to contain was to contain 2 aces. How would you modify your probability model to take into account this situation? What **assumption** underlies your modified model?

12. A lot of 20 items is known to contain 2 defectives. Consider an inspection procedure that consists of selecting 2 items at random from the shipment, one after the other, where the first item selected is not replaced before the second one is drawn. Al Williams was interested in finding the probability of the event B that the sample drawn contains 2 good items, and set up a probability model with sample space $S = \{G_1 G_2,\ G_1 D_2,\ D_1 G_2,\ D_1 D_2\}$, where $G_1 G_2$ is the event that the first and second items drawn are good, $G_1 D_2$ is the event that the first item drawn is good and the second item drawn is defective, and so on. Al assigned equal probabilities of ¼ to

these sample points and found the probability to be ¼ that the sample drawn contains 2 good items.

(a)   Is Al's conclusion valid? Explain.

(b)   State the relative-frequency interpretation of Al's conclusion.

(c)   Take 20 pennies, 2 that are new and shiny (to represent the 2 defective items in the lot) and 18 that have lost their luster (to represent the 18 good items in the lot), put them in a bag, shake the bag, and, without peeking, draw 2 pennies from the bag, one after the other. Repeat the process 300 times, record the occurrence of event B, and find the relative-frequency of B for the 300 repetitions of the process. Compare the result obtained with the relative-frequency interpretation of Al's conclusion.

(d)   Does the result obtained affect the validity of Al's conclusion? (Red Herring Alert) Explain.

(e)   Is Al's probability model realistic? Explain.

(f)   How would you set up a probability model for the selection process?

(g)   Find the probability that the sample drawn contains 2 good items in your probability model, and interpret your result in relative-frequency terms.

(h)   Do the findings obtained in (c) support the results obtained in (g)? Explain.

13. For the purpose of obtaining a maximum-likelihood estimate of the fish population of Lake Mark, 300 fish were caught, tagged, and released into the lake. Shortly thereafter, 500 fish were caught and it was found that 2 had been previously caught. Let N denotes the size of the fish population.

(a)   Setup a probability model for catching the 500 fish.

(b)   With the respect to the given conditions, what is meant by the maximum-likelihood estimate of the fish population?

(c)   Determine the maximum-likelihood function for this situation.

(d)   Find the maximum-likelihood estimate of the size of the fish population.

(e)   Is it possible that the estimate is much too low? Explain.

14. Judging from complaints about damage caused by raccoons in the community of East Beach, the raccoon population had increased significantly during the last five years along with the human population. But how large was it? This is what the community council wants to find out. They commissioned a team headed by Irving Fine to obtain a maximum likelihood estimate of the raccoon population in East Beach.

Twenty raccoons were caught, tagged and released. Shortly thereafter 15 were caught and it was found that 3 had been previously caught.

(a)   Based on these results, what is the maximum likelihood estimate of the raccoon population in East Beach?

(b)   What assumption(s) underlie this estimate?

(c)   From the damage that her property had sustained, Janet Reed felt that the estimate was too low. What aspect of the Fine team's analysis should she look over most carefully in order to satisfy herself that the estimate is realistic or present a credible case that is not? Explain.

(d)   Is it possible that the estimate is much too high? Explain.

11.5   Bernoulli Trial Modeling

It is sometimes natural and productive to view a random process as a sequence of repetitions of some basic process. For example, the tossing of a coin five times in succession may be viewed as a sequence of five repetitions of the basic process of tossing a coin once, firing at a target six times may be viewed as a sequence of six repetitions of the basic process of firing at the target, administering an antibiotic to ten people may be viewed as a ten fold repetition of the basic process of administering the antibiotic. A repetition of a basic process is called a **trial**.

A sequence of repetitions of a random process carried out under the following **conditions/assumptions** is sometimes called a **sequence of Bernoulli trials**, in honor of Jacob Bernoulli (1654-1705) who initiated a systematic study of such processes in the last part of the seventeenth century.

1. On each trial attention is focused on the occurrence or nonoccurrence of a certain event E. The occurrence of E on a given trial is called **success**; the occurrence of E^c is called **failure**.

2. The probability of success is the same for each trial; call this probability value p. The probability of failure for each trial is $1 - p$, which we shall denote by q.

3. The outcome of a trial (success or failure) **does not influence** the outcome of any other trial. That is, the outcome of any trial is independent of the outcome of any other trial.

In summary, the Bernoulli trial conditions 1, 2 and 3 lead to a class of Bernoulli trial models. To specify a model in this class we must specify the number of trials n, the event E which is the subject of our focus, and the probability p of the occurrence of E on a trial. Since the occurrence of E on a trial is called success, p expresses the probability of success on each trial. Of particular interest is the event that k successes occur in n Bernoulli trials. If X denotes the number of successes in n Bernoulli trials, then the probability of k successes is written as $P(X = k)$, which is given by:

$$P(X = k) = C(n, k)\, p^k q^{n-k}, \text{ for } k = 0, 1, 2, \ldots, n, \text{ where } q = 1 - p.$$

Students find the technical issues that emerge from the afore challenging, but even more challenging is to apply Bernoulli trial models to real world problems. What makes this harder, at least in certain problems, is that we must make a judgment call as to whether a Bernoulli trial model is a realistic fit for the random process being studied; judgment calls of this type are often difficult and may be controversial.

A Sequence of Firings at a Target

Records show that a marksman hit the bullseye two-thirds of the time. To win a prize he must hit the bullseye at least 4 times in 5 firings. What is the probability that he will win a prize?

Our first task is to set up a probability model for the process, which is to fire at the target five times. Since we have a situation which can clearly be viewed as a sequence of 5 repetitions of the basic process of firing at the target, the Bernoulli trial model defined by $n = 5$, $E =$ the event that the marksman hits the bullseye, $p = 2/3$ with $q = 1/3$, emerges as a strong possibility, depending on the **assumptions** involved.

The assignment $p = 2/3$ for the probability of success on each firing is based on long term statistical evidence of how well the marksman performed in the past. This **assumes** that the conditions under which the firings take place are uniform and that the marksman is warmed up and performing at his usual level.

As to the independence condition, we are **assuming** that the outcome of any firing has no affect, or at worst a negligible affect, on any other firing.

If we agree that these conditions are realistic, **the big IF**, then we emerge with the Bernoulli trial model defined by $n = 5$, $E =$ the event that the marksman hits the bullseye, $p = 2/3$ with $q = 1/3$, as our model for the firings. Let X denote the number of successes in 5 trials. The probability that the marksman wins a prize is $P(X \geq 4) = P(X = 4) + P(X = 5)$, which is given by:

$$C(5, 4)(2/3)^4(1/3) + C(5, 5)(2/3)^5$$
$$= 0.329 + 0.132$$
$$= 0.461$$

Larry's Apparel Shop

Records kept by Larry's Apparel Shop indicate that 20% of the shoppers who come into the store to browse actually make a purchase. What is the probability that among 100 shoppers who examine Larry's merchandise at least 30 will make a purchase?

We can identify each of the 100 shoppers with a trial, so that a sequence of 100 trials corresponding to the 100 shoppers emerges. We have $n = 100$. Take for E the event that a shopper makes a purchase. From the statistical data available on the percentage of shoppers who make a purchase, we take $p = 0.20$.

As to how **realistic** it is to take $p = 0.20$ for E for all trials, that is, for all shoppers, and to what extent the independence condition is satisfied by the trials, we should keep in mind that shoppers are sometimes acquainted and are influenced by the purchase behavior of friends or relatives. If we are willing to **assume** that such interaction can be regarded as negligible, once more the **big IF**, we are led to the Bernoulli trial model defined by $n = 100$, E = the event that a shopper makes a purchase, $p = 0.20$ with $q = 0.80$. Let X denote the number of successes (purchases made) in 100 trials. We require $P(X \geq 30) = P(X = 30) + \ldots + P(X = 100)$, which is given by:

$$C(100, 30)(0.20)^{30}(0.80)^{70} + \ldots + C(100, 100(0.20)^{100}$$

This value is approximately 0.0087 which, I emphasize to my students, gives it an aura of *precision* that we must be careful to keep in perspective, I strive to give prominence to the following perspective.

> This conclusion is **valid with respect to our model**, no-more and no-less. **If our feeling is that this model does not reflect reality sufficiently closely, then it would be foolish to regard its conclusion as realistic.**
>
> The wiser course would be to try to refine the model or develop another which better reflects reality.

The following example serves as a deterrent to the automatic pilot mode of thinking which is so easy to fall into with Bernoulli trial applications.

The Flu

Studies have led medical authorities to conclude (**assume**, if you prefer) that 1 out of 1000 people in Pittsfield has the flu. Find the probability that of 100,000 people selected 120 or more will have the flu.

We can identify each of the 100,000 people with a trial, so that a sequence of 100,000 trials emerges. We have $n = 100,000$. Let E denote the event that a person has the flu. Based on the afore studies, take $p = 0.001$, with $q = 0.999$. The Bernoulli trial model **assumption** that emerges is that whether a person has the flu or not does not have an influence on whether any other person has the flu or not.

If we accept this assumption, again the **big IF**, and let X denote the number of successes (number of people who have the flu), we have:

$$P(X \geq 120) = P(X = 120) + \ldots + P(X = 100,000)$$

$$= C(100,000;120)(0.001)^{120}(0.999)^{99,881}$$

$$+ \ldots + C(100,000;100,000)(0.001)^{100,000},$$

which can be shown to be 0.03 (rounded off).

Interesting, but let us not forget P.M.'s question to S. P. when he was told the probability that an even number showing on a toss of his die is ½, what does this mean to me? Many students are thinking along the same lines.

Interpretations of 0.03. In relative frequency terms we can say that if many groups of 100,000 are chosen (the long run) then in approximately 3% of the cases 120 or more people will have the flu. In subjective probability terms 0.03 is a measure of our degree of belief that of a group of 100,000 people chosen, 120 or more will have the flu, or equivalently in terms of odds, of 100,000 people who are chosen the odds are 1 to 32 that 120 or more people will have the flu.

All of this has an aura of *precision* which I strive to put into perspective through the following sort of observation:

> It all seems very *precise*, but appearances are often deceptive, as is the case here. This hinges on the **assumption** that whether a person has the flu or not does not have an influence on whether any other person has the flu or not.
> **Flu is highly contagious, which makes this assumption untenable.** Thus the house of mathematical cards founded on this **assumption** collapses from the point of view of realism.

11.6 Subjective Probabilities in Probability Model Terms

A number between 0 and 1, inclusive, that serves as a quantitative measure of an individual's (or group's) degree of belief in the occurrence (or non-occurrence) of an event, call it E, is called the **subjective probability of E** of the individual (or group). A strong opinion about the occurance of E is reflected by a number close to or equal to 1; a strong opinion about the non-occurance of E is reflected by a number close to or equal to 0.

Dexter White, an ardent New York Mets fan might say that the Mets have a 90% chance of taking the pennant next year—the probability that the Mets do not take the pennant next year is 0.10. Enrico Pullini, a businessman, might say that there's an 80% chance that the sale volume of my firm will top $5 million this year.—the probability that the sales volume of my firm will top $5 million this year is 0.80.

Although the term probability model is generally not used for subjective probabilities it's useful to do so because the language of modeling serves to highlight their nature as **assumptions.**

They are essentially mini probability models. Dexter White's mini probability model is P(Mets take the pennant next year) = 0.90, P(Mets do not takes the pennant next year) = 0.10; Enrico Pullini's mini probability model is P(the sale volume of my firm will top $5 million this year) = 0.80, P(the sale volume of my firm will not top $5 million next year) = 0.20.

As long as the probability model requirements are satisfied, these subjective probabilities are legitimate in a mathematical sense. How close they are to describing reality's mark is, perhaps needless-to-say, another issue.

Bill Bradley's Illness

Bill Bradley, who is suffering from a rare disease, was told by his friend Ed that, since recent medical statistics shows that 75 percent of those who have had that disease recovered, the probability that he will recover is 0.75. On the other hand, on the basis of his overall condition Dr. Aaron Sperling, who is treating Bill, believes that he has 90% chance recovery.

Who's right? Mathematically speaking, both mini probability models are *right*; reality speaking, we can't say, but Dr. Sperling gets my credibility award.

The following example is due to my colleague Dr. Irwin Kabus [3].

Subjective Probability in Banking

For the past twenty years, the top management of Morgan Guarantee Trust Company has been using a technique called histogramming to quantify and picture the uncertainty that surrounds future interest rates on which the bank's assets/liability decisions are based.

One utilizes the histogram technique by listing all the possible outcomes he believes may result from some future situation and then by assigning subjective probabilities to each of these outcomes. From observing the width of the range of outcomes and the chance associated with each a reader of the histogram is able to assess how confident the histogram maker feels about his judgments. For example, an individual's feeling about the probable level of the interest rate on a 90-day CD at some future date—say, three months from now—might look like the one shown in Figure 11.1

Each of the interest rates shown represents an interval extending 1/8% below and 1/8% above it. The percentages of subjective chances (shown by the vertical bars) are those that the individual has chosen to spread over the possible rates.

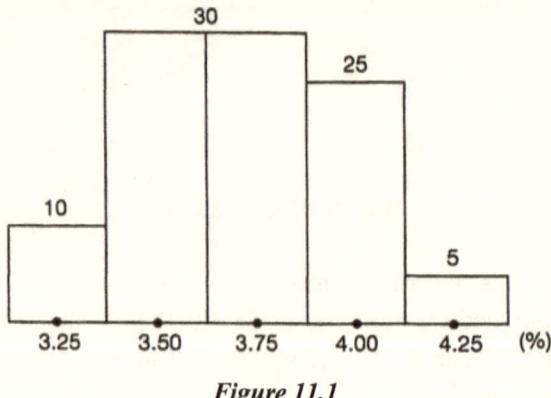

Figure 11.1

In subjective probability terms, this histogram states that the individual feels there is a 10% chance that in three months the 90-day CD rate will be between 3 & 1/8% and 3 & 3/8%, a 30% chance that it will be between 3 and 3/8% and 3 & 5/8%, a 30% chance it will be between 3 & 5/8% and 3/7%, a 25% chance it will be between 3 & 7/8% and 4 & 1/8%, and a 5% chance it will be between 4 & 1/8% and 4 & 3/8%.

This histogram indicates that the individual feels very confident that in three months the 90-day CD will be between 3 & 3/8% and 4 & 1/8%. However, he is very uncertain as to what value, within the range, the rate will take on. There are small chances that the rate could drop to 3 & 1/8% to 3 & 3/8%

or rise as high as 4 & 3/8% to 4 & 3/8%, although it is more likely that the lower rate is attained than the higher one.

As is most often case, top management does not base its decision solely on the opinion of one individual but on the opinions of several. Thus, it becomes necessary to produce an histogram whose qualitative message is some combination of the thinking of all those individuals involved in the process. Individual histograms can be combined to form a single weighted average histogram, with the weights being assigned by the individual responsible for endorsing the final histogram that top management sees.

In cases where all individual histograms receive equal weights (as is true at Morgan Guarantee) the weighted average becomes a simple arithmetic average. In other cases where the histograms are weighted differently, the weights are generally based on the qualifications and track records of the individuals producing the histograms.

At Morgan Guarantee the histogram committee consist of seven key executives from various areas of the bank. After collecting each individual's histogram, a summary sheet was generated which collectively showed the subjective probabilities assigned by each committee member. This summary serves a basis for a discussion (moderated by the analyst coordinating the histogram process) on which those members with differing views, immediately identifiable form the summary sheet, had a chance to defend them. At the end of the discussion the members of the histogram committee had the opportunity to change their histograms if they felt inclined to do so after hearing other opinions. The views of the committee as whole were then averaged into one consensus histogram which represented the views of the committee as a whole and was then presented to top management.

In modeling terms, top management was presented a mini subjective probability model that expressed the consensus of the histogram committee.

11.7   Bayes's Theorem and Math Modeling

The following is pretty much the standard approach taken to two applications of Bayes's theorem, A Reliability Question and a Marketing Problem. It is reproduced from [1; ch. 8].

On further reflection I came to appreciate that these scenarios provide an ideal setting to identify a number of **assumptions** made about matters that we tend to take for granted, but should be identified as **assumptions**. Once they are identified as such the door is opened to give thought to their relative importance in the scheme of things.

Version 1, [1; ch. 8]

A Reliability Question

The following data are available on the reliability of a blood test for detecting the presence of a certain disease. Of people with the disease, 90 percent of the blood test examinations detected the disease and 10 percent went undetected. Of the people who did not have the disease, 98 percent of those given the blood test were correctly diagnosed as not having the disease, and 2 percent were incorrectly diagnosed as having the disease. In a certain population it is estimated that 2 percent have the disease. A person is selected at random from the population. If, on the basis of the blood test, it is concluded that he has the disease, what is the probability that he actually has the disease?

To analyze this problem, we must be very careful to properly identify the event whose occurrence is given (designated by B), and the event whose probability we are seeking, given the occurrence of B. For this situation, we introduce B, A_1, and A_2 as follows:

B: the blood test indicates the presence of the disease.

A_1:   a person with the disease is selected.

A_2:   a person not having the disease is selected.

The problem is to determine $P(A_1/B)$, the probability that the person selected has the disease given that the blood test indicates the presence of the disease. For this situation we have the following background:

$P(A_1)$, the probability that the person selected has the disease.

$P(A_2)$, the probability that the person selected does not have the disease.

$P(B/A_1)$, the probability that the blood test indicates the presence of the disease given that the person selected has the disease.

$P(B/A_2)$, the probability that the blood test indicates the presence of the disease given that the person selected does not have the disease.

The randomness of the selection and the data given lead us to the following assignment of probability values.

$P(A_1) = 0.02$ (2% of the population are estimated to have the disease.)

$P(A_2) = 0.98$ (98% of the population do not have the disease.)

$P(B/A_1) = 0.9$ (Of those with the disease, 90% were correctly diagnosed as having the disease.)

$P(B/A_2) = 0.02$ (Of those not having the disease, 2% were incorrectly diagnosed as having the disease.)

Figure 11.2 gives us a probability-tree view of this situation.

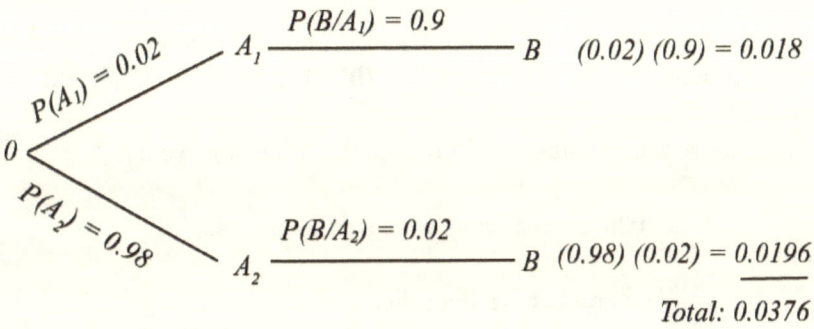

Figure 11.2

From Bayes's theorem we obtain:

$$P(A_1 / B) = \frac{0.018}{0.0376} = 0.48$$

The relative-frequency interpretation of this result is that, over the long run, of the people selected at random and diagnosed as having the disease, less than one half (approximately 48 percent) will actually have the disease.

Version 1 [1; ch.8]

A Marketing Problem

In response to an antipollution drive the Starr Company is planning to market a new cleaning product that does not exhibit pollution side effects. The marketing department of the Starr Company initially assigned a probability of 0.5 to the product being a big seller, a probability of 0.3 to the product being a fair seller, and a probability of 0.2 to the product being a poor seller. A marketing test was planned and carried out. If the product is a big seller, it is estimated that the probability of selling 7000 or more units in the test is 0.6. If the product is a fair seller, it is estimated that the probability of selling 7000 or more units in the test is 0.9. If the product is a poor seller, the probability of selling 7000 or more units is estimated to be 0.2.

If 9000 units were sold in the marketing test, what is the probability that the cleaning product will be (a) a fair seller; (b) a big seller; (c) a fair seller or big seller?

To analyze this problem, we introduce the following events:

B: 7000 or more units were sold in the test.

A_1:   the product is a big seller.

A_2:   the product is a fair seller.

A_3:   the product is a poor seller.

The problem is to determine $P(A_2/B)$, the probability that the product is a fair seller given that 7000 or more units were sold in the test; $P(A_1/B)$, the probability that the product is a big seller given that 7000 or more units were sold in the test. For this situation we have the following background:

$P(A_1)$, the probability that the product is a big seller.

$P(A_2)$, the probability that the product is a fair seller.

$P(A_3)$, the probability that the product is a poor seller.

$P(B/A_1)$, the probability that 7000 or more units are sold in the test given that the product is a big seller.

$P(B/A_2)$, the probability that 7000 or more units are sold in the test given that the product is a fair seller.

$P(B/A_3)$, the probability that 7000 or more units are sold in the test given that the product is a poor seller.

The given **assumptions** made by the marketing department lead us to the following assignment of probability values:

$$P(A_1) = 0.5, \qquad P(A_2) = 0.3 \qquad P(A_3) = 0.2$$
$$P(B/A_1) = 0.6 \qquad P(B/A_2) = 0.9 \qquad P(B/A_3) = 0.2$$

A probability-tree view of this situation is provided by Figure 11.3.

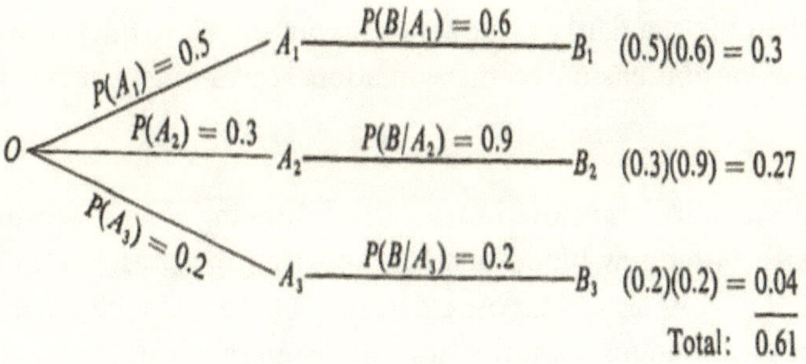

Figure 11.3

From Bayes's theorem we have:

(a) $P(A_2/B) = \dfrac{0.27}{0.61} = 0.44$

(b) $P(A_1/B) = \dfrac{0.3}{0.61} = 0.49$

(c)   The probability of a fair or big seller, given that 7000 or more units were sold in the test is $0.44 + 0.49 = 0.93$

Considering the once-and-only nature of this situation, the probability assignments made and the results obtained are to be interpreted in subjective terms. The value 0.93 for the probability of a fair or big seller is a valid conclusion of initial judgments translated into quantitative terms [$P(A_1) = 0.5$, $P(B/A_1) = 0.6$, and so on **assumptions**], and may be helpful to the management of the Starr Company for making a decision on whether or not to market the new cleaning product on a major scale. It is important to keep in mind that the value 0.93, as a quantitative expression of the product's potential as a fair or big seller, does not go beyond the initial inputs. Never forget the NINO principle; if nonsense in, nonsense out.

Version 2 [2; ch. 8]

A Reliability Question

A new test has been developed for determining whether a person has AIDS (or cancer or any number of medical conditions). It is simpler and less expensive to apply than the standard test now in use, but there are questions about its reliability. One dimension of this situation is examined in the following example.

Against a standard test now in use, the following data are available on the reliability of a new blood test for detecting the presence of a certain disease (fill in your not-so-favorite disease). Of people with the disease, 90 percent of the blood test examinations detected the disease and 10 percent went undetected. Of the people who did not have the disease, 98 percent of those given the blood test were correctly diagnosed as not having the disease, and 2 percent were incorrectly diagnosed as having the disease. In a certain population it is estimated that 2 percent have the disease.

A person is selected at random from the population. If, on the basis of the blood test, it is concluded that he has the disease, what is the probability that he actually has the disease (P (person +/test +))?

To analyze this problem, we must be very careful to properly identify the event whose occurrence is given (designated by B), and the event whose probability we are seeking, given the occurrence of B. For this situation, we introduce B, A_1, and A_2 as follows:

B: the blood test indicates the presence of the disease (test is +).

A_1:   a person with the disease is selected (person is +).

A_2:   a person not having the disease is selected (person is -).

The problem is to determine $P(A_1 / B) = P$(person + /test+), the probability that the person selected has the disease given that the blood test indicates the presence of the disease.

For this situation we have the following background:

$P(A_1)$, the probability that the person selected has the disease (person is +).

$P(A_2)$, the probability that the person selected does not have the disease (person is -).

$P(B / A_1)$, the probability that the blood test indicates the presence of the disease given that the person selected has the disease (P (test +/ person +)).

$P(B / A_2)$, the probability that the blood test indicates the presence of the disease given that the person selected does not have the disease (P (test +/ person -)).

The randomness of the selection and the **assumptions** given lead us to the following assignment of probability values.

$P(A_1) = 0.02$ (2% of the population are estimated to have the disease.)

$P(A_2) = 0.98$ (98% of the population do not have the disease.)

$P(B / A_1) = 0.9$ (Of those who are +, 90% tested +)

$P(B / A_2) = 0.02$ (Of those who are -, 2% tested +))

Figure 11.4 gives us a probability-tree view of this situation.

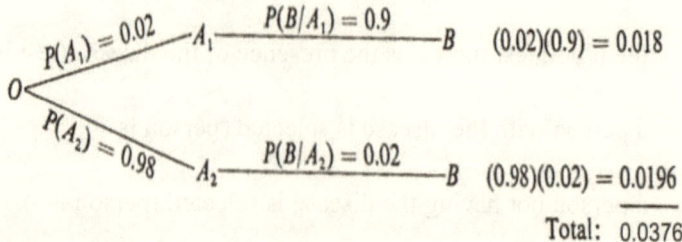

Figure 11.4

From Bayes's theorem we obtain:

$$P(A_1 / B) = \frac{0.018}{0.0376} = 0.48$$

The relative-frequency interpretation of this result is that of many people selected at random and diagnosed as + less than one half (approximately 48%) will be +.

In terms of odds, if a person chosen at random is diagnosed as + on the basis of this blood test, the odds that he/she is + is 0.48 to 0.52 or, equivalently, 12 to 13. In terms of what would be considered a fair bet, it's my $12 against your $13 that he/she is +.

Beware the Assumptions

As *precise* as P(person is +/test is +) = 0.48 is, mathematically speaking, its realism hinges on the realism of the **assumptions** made:

1. For the population being considered it is estimated that 2% have the disease, which raises the question, how reliable is this estimate?

2. Of those who are + (in terms of the standard test), 90% of the blood test examinations were +, while raises a question about the reliability of the standard test.

3. Of those who were—(in terms of the standard test), 98% of those given the blood test were correctly diagnosed as -, which again raises a question about the reliability of the standard test.

Postulate P: Other factors that may bear on the reliability of the blood test (such as the level of competence of those administering the test and the quality of the test materials) are considered negligible.

Version 2 [2; ch. 8]

Marketing

Marketing is of vital concern to every company that produces a product. The finest product imaginable if not known to its potential consumers is, as the saying goes, dead in the water.

We consider the case of a company whose marketing department has made initial judgments about their product being a big, fair or poor seller which they have expressed in quantitative form as subjective probabilities. To refine these initial judgments they plan to carry out a limited scale marketing test. Bayes's theorem can be employed to move from the prior probabilities of their initial judgments to after-the-test posterior probabilities.

In response to an antipollution drive the Starr Company is planning to market a new cleaning product that does not exhibit pollution side effects. The marketing department of the Starr Company initially assigned a probability of 0.5 to the product being a big seller, a probability of 0.3 to the product being a fair seller, and a probability of 0.2 to the product being a poor seller. A marketing test was planned and carried out. If the product is a big seller, it is **estimated (assumed**, if you prefer) that the probability of selling 7000 or more units in the test is 0.6. If the product is a fair seller, it is **estimated** that the probability of selling 7000 or more units in the test is 0.9. If the product is a poor seller, the probability of selling 7000 or more units is **estimated** to be 0.2.

If 9000 units were sold in the marketing test, what is the probability that the cleaning product will be (a) a fair seller; (b) a big seller; (c) a fair seller or big seller?

To analyze this problem, we introduce the following events:

B: 7000 or more units were sold in the test.

A_1:   the product is a big seller.

A_2:   the product is a fair seller.

A_3:   the product is a poor seller.

The problem is to determine $P(A_2 / B)$, the probability that the product is a fair seller given that 7000 or more units were sold in the test; $P(A_1 / B)$, the probability that the product is a big seller given that 7000 or more units were sold in the test. For this situation we have the following background:

$P(A_1)$, the probability that the product is a big seller.

$P(A_2)$, the probability that the product is a fair seller.

$P(A_3)$, the probability that the product is a poor seller.

$P(B / A_1)$, the probability that 7000 or more units are sold in the test given that the product is a big seller.

$P(B / A_2)$, the probability that 7000 or more units are sold in the test given that the product is a fair seller.

$P(B/A_3)$, the probability that 7000 or more units are sold in the test given that the product is a poor seller.

The **assumptions** made lead us to the following assignment of probability values:

$$P(A_1) = 0.5, \qquad P(A_2) = 0.3 \qquad P(A_3) = 0.2$$

$$P(B/A_1) = 0.6 \qquad P(B/A_2) = 0.9 \qquad P(B/A_3) = 0.2$$

A probability-tree view of this situation is provided by Figure 11.5.

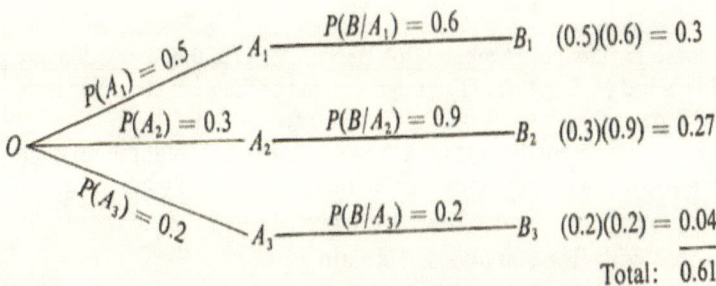

Figure 11.5

From Bayes's theorem we have:

(a) $P(A_2/B) = \dfrac{0.27}{0.61} = 0.44$

(b) $P(A_1/B) = \dfrac{0.3}{0.61} = 0.49$

(c)   The probability of a fair or big seller, given that 7000 or more units were sold in the test is $0.44 + 0.49 = 0.93$

We also have that $P(A_3/B) = 0.07$.

In summary, we have Table 11.3.

Table 11.3

	Prior probability	Posterior probability given that 7000 or more units were sold
A_1: big seller	0.5	0.49
A_2: fair seller	0.3	0.44
A_3: poor seller	0.2	0.07

This scenario provides us with a good opportunity to address a math misunderstanding that is far too often believed.

Beware the Assumptions

"It doesn't make sense", argued Bob Benson, CEO of the Starr Company to his board of directors. "The requirements of the marketing test were met, and yet the probability of a big seller went down.
It's Bayes's theorem that's at fault. You sold me on this highfalutin math and now we have math nonsense. No more highfalutin math for me."

Do you agree with Bob's analysis? Explain.

In all honesty you would have to take issue with Bob's analysis. You would want to point out that "Bayes's theorem did not generate math nonsense; it delivered what it is capable of—valid conclusions based on **our assumptions**.

We should carefully review our initial assignment of subjective probabilities. There is where the fault, if there is fault, is to be found."

Whether honesty is the best policy or not depends on circumstances. It may get you fired or it may get you promoted.

11.8   Food for Thought Questions

Set up Bernoulli trial models for the situations described in 15-17, state the underlying **assumptions** and comment on their realism, and determine or state (in combinatorial form) the probabilities asked for. In each case express a judgment on the suitability of a Bernoulli trial model for the process in question.

Is a more realistic model available for any of the situations described? If so, state the model, comment on its realism, and determine the probabilities required.

15. Consider an examination that consists of 10 multiple-choice questions. Each question allows four choices, with only one answer being correct. If a student guesses the answer for each question, what is the probability that he will pass the exam (6 or more correct answers)?

16. According to company records, 1% of the output of light bulbs in a certain mass production process are defective. What is the probability that of 1,000 light bulbs produced, no more than 15 are defective?

17. The fruit fly *Drosophila* has been widely used for genetic studies and has furnished much of the evidence for modern genetic theory. It has been shown that mutations can be induced in *Drosophila* by subjecting the flies to X-rays, ultraviolet radiation, or any high-energy radiation, as well as chemical compounds such as those in the mustard gas family. Evidence suggests that 0.01% of large batches of *Drosophila* subjected to high-energy radiation develop a mutation.

    What is the probability that of 20,000 *Drosophila* subjected to high-energy radiation, at least one will develop a mutation?

18. A recently developed antibiotic has been found to cause an allergic reaction in one out of 200 people injected. Find the probability that fewer than 100 reactions occur among 10,000 people injected with the antibiotic.

19. Studies have led medical authorities to conclude that 1 out of 500 teenagers in Southchester has the measles.

    (a)   Find the probability that out of 10,000 teenagers selected at most 25 will have the measles.

(b)   Interpret the result you obtained.

(c)   Are the numbers obtained in answer to (a) and (b) credible? Explain.

Bayes's Theorem

20. In connection with the discussion of a reliability question, suppose that a person is selected at random from the population and, on the basis of the blood test, it is concluded that he does not have the disease.

(a)   What is the probability that he has the disease?

(b)   What is the relative-frequency interpretation of your answer?

(c)   What **assumptions** underlie the conclusion and its interpretation obtained in (a) and (b)?

(d)   Does the application of Bayes's theorem prove that the estimate obtained from (a) and (b) is *correct*? Explain.

(e)   Is it possible that the **estimate** (valid conclusion) is much too low? Explain.

21. The following data are available on the reliability of a certain X-ray detection procedure for lung cancer. Of those people with lung cancer who were tested, 98 percent of the cases were detected, whereas 2 percent were undetected; 99 percent of those who did not have lung cancer who were tested were correctly diagnosed as not having lung cancer, whereas 1 percent were misdiagnosed.

    In a certain highly populated industrial center it is **estimated (assumed**, if you prefer) that 3 percent have lung cancer.
A person is selected at random from the population and tested.

(a)   If the person is diagnosed as having lung cancer, what is the probability that he has the disease?

(b)   If the person is diagnosed as not having lung cancer, what is the probability that he has the disease?

(c)   What are the relative-frequency interpretations of your conclusions?

(d)   What **assumptions** underlie the conclusions and their interpretations obtained in (a), (b), and (c)?

(e)   Does the application of Bayes's theorem prove that the estimates obtained from (a), (b), and (c) are *correct*? Explain.

(f)   Is it possible that the **estimates** obtained from (a), (b), and (c) are much too high? Explain.

22. The following data are available on the reliability of a test to determine when the level of mercury contamination in fish exceeds what is considered to be a permissible level. Of the fish with excess mercury contamination that were tested, 99 percent of the cases were detected, whereas 1 percent went undetected; 96 percent of the fish that were tested whose mercury content did not exceed the permissible level were correctly diagnosed as being within the permissible level, whereas 4 percent were misdiagnosed.

For Lake Bennett it is **estimated** that 6 percent of the fish population have a mercury content that exceeds the permissible level.

A fish is caught from the lake and tested.

(a)   If the mercury content of the fish is diagnosed as excessive, what is the probability that it is not excessive?

(b)   If the mercury content of the fish is diagnosed as within the permissible level, what is the probability that it is excessive?

(c)   What are the relative-frequency interpretations of your conclusions?

(d)   What **assumptions** underlie the conclusions and their interpretations obtained in (a), (b), and (c)?

(e)   Does the application of Bayes's theorem prove that the **estimates** obtained in (a) and (b) are *correct*? Explain.

(f)   Is it possible that the results obtained in (a), (b), and (c) are much too low? Explain.

Mercury contamination of large ocean fish such as sword and tuna fish is a problem of great concern. How reliable are "safe" exposure levels? It is noteworthy that the Food and Drug Administration "safe" exposure level to methymercury in fish (0.4 micrograms per kilogram of a person's body weight per day) has been criticized as misleading. J. Gorman, "Does Mercury Matter? Experts Debate the Big Fish Question", *The New York Times*, July 29, 2003; D5.

23. In connection with the discussion of the Starr Company's marketing problem, given that 9000 units were sold in the marketing test, what is the probability that the cleaning product is a poor seller?

24. The Borg Company is planning to market an inexpensive computer especially designed for classroom use. Their sales department initially **estimated** that the probability of a big seller is 0.5, the probability of a good seller is 0.3, and the probability of a poor seller is 0.2. A market test was planned and carried out. It was estimated by the sales department that, if the computer is a big seller, the probability of selling more than 200 units in the test is 0.9; if the computer is a good seller, it was estimated that the probability of selling more than 200 units in the test is 0.5; and if the computer is a poor seller, it was estimated that the probability of selling more than 200 units in the test is 0.1.

If more than 200 units were sold in the marketing test, what is the probability that:

(a)   the computer is a big seller;

(b)   the computer is a good seller;

(c)   the computer is a poor seller;

(d)   the computer is a big or good seller?

(e)   If the conclusions obtained in (a), (b), (c) and (d) turn out to be unrealistic, is the source of the difficulty Bayes's theorem? Explain.

(f)   Is it possible that the results obtained in (a), (b), and (c) much too low? Explain.

Subjective Probability and Relative Frequency Interpretations of Probability

25. It is asserted that the probability of an event E is 0.80. State the relative-frequency and subjective-probability interpretations of this assertion, and describe the main features of these interpretations.

26. After watching a pair of dice being tossed 100 times, an observer commented that the probability of an even sum showing on the 101st toss is 0.85.

   Is this probability assignment one that is to be interpreted in relative-frequency or subjective-probability terms? Explain.

27. The following comment appeared in an article on natural gas supplies (*New York Times*, Feb. 22, 1977, p. 14): "How much gas is left to be discovered? . . . The last Geological Survey estimate, made two years ago, was this: Given available technology and current economics, there is a 95 percent probability that 322,000 billion cubic feet can be located and produced; there is a 5 percent probability that 655,000 billion cubic feet can be located and produced."

   How should these probabilistic statements be interpreted? Explain.

28. In a letter to the editor of the *New York Times* (Feb. 28, 1971) on the background of the atomic bomb, Hans Bethe wrote, "By February 1945 it appeared to me and to other fully informed scientists that there was a better than 90 per cent probability that the atomic bomb would in fact explode . . . ."

   How should this probabilistic statement be interpreted? Explain.

29. Eric Roberts, a student at Huxley College, commented that the probability that he will get an A in Sociology this semester is 0.95.

     Is this probability value to be interpreted in relative frequency or subjective probability terms. How so?

30. On Dec. 29, 1978 two acoustics experts said tests showed a probability of 95 percent or better that a shot was fired from a grassy knoll in Dallas where President John F. Kennedy was assassinated. This testimony was presented before the House Select Committee on Assassinations. (*New York Times*, Dec. 30, 1978).

     How should this probability statement be interpreted? Explain.

11.9   Watch Your Language: Normally Distributed vs. Approximately Normally Distributed

Preface

Sound familiar? A line is the shortest distance between two points. The correct version is, A line is the path of shortest distance between two points (at least on a flat surface). Sound familiar? Consider the line $2x + 3y = 6$. The correct version is, consider the line whose equation is $2x + 3y = 6$. We agree to such "abuses" of language, shortcuts if you prefer, to avoid language from becoming too cumbersome.

The key prerequisite for their use is discretion about the line which distinguishes acceptable linguistic shortcut from a statement that's just plain wrong. Describing π, the ratio of the circumference of a circle to its diameter, as 3.14 is just plain wrong, no ifs-ands-or-buts about it, and this characterization of π is bound to sow confusion and misunderstanding. π is approximately 3.14 (correct to two places) and the qualifier approximately makes all the difference in the world.

The Point of the Afore Preface

A statistics book with many admirable qualities that I refer to from time to time describes the normal curve as a theoretical distribution, which of course it is. The condition that a random variable be normally distributed is a mathematical ideal needed as a standard for proving theorems, but this ideal is not "exactly" realized in practice.

Yet, concerning an application the afore book reads, "The attendance of an athletic stadium is normally distributed with mean . . . . Find . . . ." This should read, The attendance at an athletic stadium is *approximately* normally distributed . . . . To omit *approximately* is to equate the normal curve math model ideal with a less than ideal real world situation. This, I submit is an abuse of language that crosses the line between acceptable linguistic shortcut and no ifs-ands-or-buts wrongheaded and misleading. Alas, many statistics books that I had occasion to look at are guilty of the same practice. **Beware.**

11.10   References of Interest?

1.   W. J. Adams, *Finite Mathematics, Models, and Structure*, Revised Printing (Dubuque: Kendall/Hunt Pub. Co., 1999).

2.   W. J. Adams, *Finite Mathematics, Models, and Structure*, Revised Edition (Philadelphia: Xlibris, 2009). This book is available on the web at webpage.pace.edu/wadams.

Concerning Subjective Probability

3.   I. Kabus, "You can Bank on Uncertainty," *Harvard Business Review*, 54, 3 (May-June 1976), 95-105.

4.   L. J. Savage, *The Foundations of Statistics* (New York: Wiley and Sons, Inc., June 1976), 95-105.

5.   R. Schlaifer, *Probability and Statistics for Business Decisions* (New York: McGraw Hill, 1959).

6.   R. Schlaifer, *Analysis of Decisions Under Uncertainty* (New York: McGraw Hill, 1969).

12

Math Modeling for Conflict Situations

12.1 Preface

In the mathematical theory of games there is envisioned two or more parties (called **players**) with conflicting interests who have a certain freedom of choice in the selection of options, but whose control over the situation is partial at best. A player has control over his own actions within the rules that define the game, but does not have control over his opponents' actions, nor does he have control over the element of chance if it is present. The problem is to define, within the rules that govern the game, a concept of optimal strategies for the players and develop methods for determining these optimal strategies. If the number of opponents is n, we speak of an **n-person game**. If whatever any player wins at the end of the game is lost by other players, so that the sum of the payoffs to all players is zero (winnings are denoted by positive payoffs and losses by negative payoffs), then the game is called a **zero-sum game**.

Since I believe it reasonable to assume that game theory was on the periphery of the math studies of most of us, some introductory comments might be useful.

12.2 Introduction to Game Theory

In this introduction I restrict our attention to a special class of two-person zero-sum games, the games of matrix type.

To illustrate a matrix-type two-person zero-sum game, consider the following situation. There are two opponents, Mr. Row and Mr. Column, and a rectangular array of numbers

$$
\begin{array}{c}
\begin{array}{ccc} C_1 & C_2 & C_3 \end{array} \\
\begin{array}{c} R_1 \\ R_2 \end{array}
\begin{bmatrix} 2 & -2 & -3 \\ 3 & 1 & 2 \end{bmatrix}
\end{array}
$$

which contains two rows R_1 and R_2 and three columns C_1, C_2, and C_3.

Mr. Row is to choose a row and Mr. Column is to choose a column. Neither player knows the other player's choice. When the choices made are revealed, the payoff to each player is specified by the entry in the above array in the intersection of the row and column chosen. Thus if Mr. Row chooses row R_1 and Mr. Column chooses column C_1, the payment of Mr. Column to Mr. Row is 2 units, 2 dollars, let us say. (The units may be monetary or of some other nature.) If Mr. Row chooses row R_1 and Mr. Column chooses column C_2, the payment of Mr. Column to Mr. Row is—$2, which means that Mr. Row pays Mr. Column $2. Since the game is a zero-sum game, the amount won by one player is equal to the amount lost by the other player. The rectangular array of numbers that specifies the payoffs is called the **payoff matrix to the row player**.

The problem is to determine each player's **best option** so as to guarantee the best possible outcome under the conditions that define the game.

More generally, a **two-person zero-sum game of matrix type** satisfies the following conditions:

1.　There are two opponents, called the **row and column players**.

2.　A payoff matrix describing the payoffs from the column player is defined and is known to both players. The amount won by one player is equal to the amount lost by the other player so that the sum of the payoffs is zero.

3.　Each player makes one move (a row is chosen by the row player and a column is chosen by the column player) without knowing his opponent's choice.

4.　The payoff from the column player to the row player is described by the value in the payoff matrix that is in the intersection of the row and column chosen.

5.　It is **assumed** that both players will do the best they can within the restrictions of the game.

The problem is to define and determine optimal strategies for the row and column players within the context imposed.

Optimal Pure Strategies

Returning to the adventures of Mr. Row and Mr. Column, the question is: what choice should each player elect to guarantee himself the best possible outcome no matter what his opponent does?

Looking at the situation from Mr. Row's point of view, we see that if Mr. Row chooses R_1, the worst that can happen is that Mr. Column may pick C_3, in which case Mr. Row loses \$3. If Mr. Row chooses R_2, the worst that can happen is that Mr. Column may pick C_2, in which case Mr. Row wins \$1. Thus the safe choice for Mr. Row is row R_2, since he is guaranteed to win \$1 even if Mr. Column makes his best choice, and he may do better if Mr. Column makes a poor choice. Thus Mr. Row looks at the minimum payoff offered by each row, called the **row minima**, and chooses the row, R_2 in this case, that yields the maximum of these row minima, called **maximin**. Thus maximin = 1.

Looking at the situation from Mr. Column's point of view, we see that if Mr. Column chooses C_1, the worst that can happen is that Mr. Row may choose R_2, in which case Mr. Column loses \$3. If Mr. Column chooses C_2, the worst that can happen is that Mr. Row may choose R_2, in which case Mr. Column loses \$1. If Mr. Column chooses C_3, the worst that can happen is that Mr. Row may choose R_2, in which case Mr. Column loses \$2. Thus the safe choice for Mr. Column is C_2, since he is guaranteed that Mr. Row can win no more than \$1 even if Mr. Row makes his best choice, and he may do better if Mr. Row makes a poor choice. Thus Mr. Column looks at the maximum payoff in each column, called the **column maxima**, and chooses the column, C_2 in this case, that yields the minimum of these row maxima, called **minimax**. Thus minimax = 1.

In this situation maximin = minimax = 1. The results of this analysis are summarized in Figure 12.1. Row minima are set down next

		Mr. Column				
		C_1	C_2	C_3		Row minima
Mr.	R_1	2	−2	−3		−3
Row	R_2	3	1	2		1
						Maximin = 1
Column maxima		3	1	2		Minimax = 1

Figure 12.1

to each row, column maxima are set down below each column, and maximin and minimax are recorded as shown.

> More generally, if maximin = minimax, the choice of the row that yields this common value is called an **optimal pure strategy for the row player**, and the choice of the column that yields this common value is called an **optimal pure strategy for the column player**.
>
> The common value of maximin and minimax is called the **value** of the game. A **fair game** is one with value 0.

The Asta Company vs. the Audre Company

The Asta and Audre companies are competing for shares in the estimated $5 million minicomputer market. If, it is **assumed**, both companies conduct high-intensity advertising campaigns, they will each get 50 percent of the market. If, it is **assumed,** Asta runs a high-intensity campaign to Audre's low-intensity campaign, Asta will capture 70 percent of the market (to Audre's 30 percent); if, it is **assumed**, Asta runs a low-intensity campaign against Audre's high-intensity blitz, Asta will capture 40 percent of the market (to Audre's 60 percent). If, it is **assumed,** both run low-intensity campaigns, Asta will capture 65 percent of the market (to Audre's 35 percent).

With respect to these assumptions, determine each company's optimal strategy.

The outcome resulting from the various options can be expressed in terms of the game defined by the payoff matrix shown in Figure 12.2, where the numbers given express the percentages of the market Audre concedes to Asta.

$$
\begin{array}{cc}
 & \text{Audre} \\
 & \begin{array}{cc} \text{Low} & \text{High} \end{array} \\
\text{Asta} \quad \begin{array}{c} \text{Low} \\ \text{High} \end{array} & \begin{bmatrix} 0.65 & 0.40 \\ 0.70 & 0.50 \end{bmatrix}
\end{array}
$$

Figure 12.2

Thus if Asta chooses row 1 (opts for a high-intensity campaign), while Audre chooses column 1 (opts for a low-intensity campaign), Asta will capture 65 percent of the market.

The row minima and column maxima are summarized in Figure 12.3, from which we see that maximin = minimax = 0.50. Thus it follows as a valid conclusion

		Audre		
		Low	High	Row minima
Asta	Low	0.65	0.40	0.40
	High	0.70	0.50	0.50
				Maximin = 0.50
Column maxima		0.70	0.50	Minimax = 0.50

Figure 12.3

that the Asta Company's optimal strategy is to play row 2 (opt for a high-intensity campaign), and the Audre Company's optimal strategy is to play column 2 (opt for a high-intensity campaign). The anticipated outcome is that both companies will obtain an equal share of the market.

Should the Asta and Andre Companies Implement these Findings?

If the companies implement these findings will the anticipated outcome come to pass? Here too, we can only note that game theory, like every mathematical discipline, **guarantees the validity** of its conclusions, **not their truth/realism**. The valid conclusions are realistic if the underlying **assumptions** are sufficiently realistic, and thus both companies would be well advised to review the **assumptions** made.

Is it sufficiently **realistic** to reduce this competitive situation to the game-theory terms that have been used, or are there some basic features that have not been adequately taken into account? This is the key consideration.

Mixed Strategies

Maximin is not equal to minimax for all matrix games, and thus the approach used to define optimal strategies is of limited applicability. To illustrate, consider the following version of the game of matching pennies. Two players each show head or tail without knowing the other player's choice. If they both show heads, the row player pays the column player 4¢, and if they both show tails the row player pays the column player 3¢. If the row player shows heads and the column player show tails, the column player pays the row player 5¢; if the row player shows tails and the column player shows heads, the column player pays the row player 2¢. The payoff matrix for the row player is shown in Figure 12.4, from which we see that maximin = -3, while minimax = 2.

	H	T	Row minima
H	-4	5	-4
T	2	-3	-3
			Maximin = -3
Column maxima	2	5	Minimax = 2

Figure 12.4

It's a matter of chance as to who wins on any one play of the game, and the analysis that suffices when maximin equals minima is insufficient here.

Let us imagine the game being played a large number of times, and try to develop an optimal long-run strategy for playing the game. If the row player plays row 1 (shows heads) all the time, the column player would react by playing column 1 (showing heads) and thus winning 4¢ on every play. A similar situation arises if row 2 is played all the time. Obviously, the row player should play row 1 a certain percentage of the time and row 2 the rest of the time, and in a random way so that no pattern can be discerned and taken advantage of.

At this point it is useful to introduce the language of probability. To say, for example, that row 1 should be played with probability $\frac{1}{3}$ and row 2 should be played with probability $\frac{2}{3}$ is interpreted to mean that row 1 should be played roughly one third of the time and row 2 should be played two thirds of the time in a random way.

> More generally, to say that row 1 should be played with probability p and row 2 should be played with probability $1 - p$ is interpreted to mean that row 1 should be played roughly p of the time and that row 2 should be played the rest, or $1 - p$, of the time.
> The question is, what is the **best value** of p?

To answer this question, we must introduce the idea of expected value.

The **row player's expected value with respect to column 1 being played by the column player** is the sum of the products of the payoffs in column 1 times the probabilities of these payoffs. Since the payoffs in column 1 are -4 and 2 (see Figure 12.5),

Play row 1 with probability p $\qquad\qquad \begin{bmatrix} -4 & 5 \\ 2 & -3 \end{bmatrix}$

Play row 2 with probability 1 - p

Figure 12.5

and the probabilities with which they occur are the probabilities with which row 1 and row 2 are chosen, that is, p and $1 - p$, respectively, the row player's expected value with respect to column 1 being played by the column player is

$$y_1 = (-4)p + 2(1 - p)$$
$$= -6p + 2, \qquad \text{where } 0 \leq p \leq 1$$

Similarly, **the row player's expected value with respect to column 2 being played by the column player** is defined as the sum of the products of the payoffs in column 2 times the probabilities of these payoffs. Since the payoffs in column 2 are 5 and -3, and the probabilities with which they occur are p and $1 - p$, respectively, the row player's expected value with respect to column 2 being played by the column player is

$$y_2 = 5p + (-3)(1 - p)$$
$$= 8p + 3, \qquad \text{where } 0 \leq p \leq 1$$

> Expected value y_1 can be interpreted as approximating the average amount won per game by the row player with respect to column 1 being played by the column player, **assuming** a large number of plays of the game, as the following analysis shows. By definition,

$$y_1 = (-4)p + 2(1 - p).$$

Since p is roughly the fraction of the time row 1 is played, and $1 - p$ is the fraction of the time row 2 is played, y_1 is approximately given by:

$$y_1 = (-4)\frac{\left[\begin{array}{c} no.\,of\,times \\ row\,1\,is\,played \end{array}\right]}{\left[\begin{array}{c} no.\,of\,times \\ the\,game\,is \\ played \end{array}\right]} + (2)\frac{\left[\begin{array}{c} no.\,of\,times \\ row\,2\,is\,played \end{array}\right]}{\left[\begin{array}{c} no.\,of\,times \\ the\,game\,is \\ played \end{array}\right]}$$

Since the number of times the game is played is a common denominator, we can add numerators to obtain:

$$y_1 \approx \frac{(-4)\begin{bmatrix} no.\,of\,times \\ row\,1\,is\,played \end{bmatrix} + (2)\begin{bmatrix} no.\,of\,times \\ row\,2\,is\,played \end{bmatrix}}{\begin{bmatrix} no.\,of\,times\,the\,game\,is\,played \end{bmatrix}}$$

Since -4 is the amount won by the row player when row 1 is played and 2 is the amount won by the row player when row 2 is played, we have:

$$(-4)\begin{bmatrix} no.\,of\,times \\ row\,1\,is\,played \end{bmatrix} = \begin{bmatrix} row\,player's\,winnings \\ when\,row\,1\,is\,played \end{bmatrix}$$

$$(2)\begin{bmatrix} no.\,of\,times \\ row\,2\,is\,played \end{bmatrix} = \begin{bmatrix} row\,player's\,winnings] \\ when\,row\,2\,is\,played \end{bmatrix}$$

The sum of the winnings with respect to rows 1 and 2 being played is the row player's total winnings. Thus we obtain

$$y_1 \approx \frac{\begin{bmatrix} row\,player's\,total\,winnings \end{bmatrix}}{\begin{bmatrix} no.\,of\,times\,the\,game\,is\,played \end{bmatrix}}$$

which is the row player's average winnings per game with respect to the column player playing column 1.

Similarly, the expected value y_2 approximates the row player's average winnings per game with respect to the column player playing column 2, **assuming** a large number of plays of the game.

We want to choose p so as to maximize, in some sense, the row player's expected average winnings per game. This idea can be precisely expressed in the following way.

Let:

$$m = \text{minimum of } y_1 \text{ and } y_2$$

Choose p so as to maximize m. That is, choose p so as to maximize the minimum of the expected values with respect to columns 1 and 2 being played by the column player.

To determine p we turn to the graphs of y_1 and y_2 shown in Figure 12.6. From the graphs of y_1 and y_2 we obtain the graph of $m = $ minimum of y_1 and y_2, shown in Figure 12.7.

Setting y_1 equal to y_2 and solving for p yields:

$$-6p + 2 = 8p - 3$$
$$-14p = -5$$
$$p = \frac{5}{14}$$

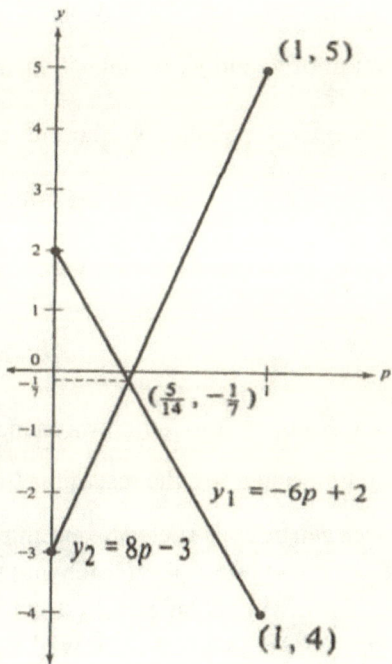

Figure 12.6

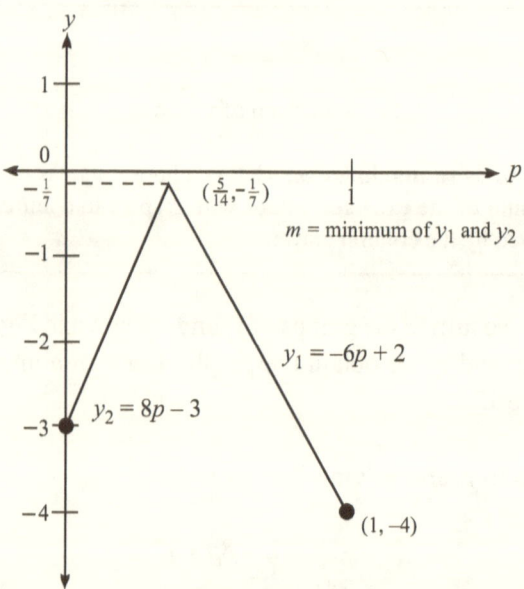

Figure 12.7

Substituting $\frac{5}{14}$ for p in y_1 or y_2 yields $y_1 = y_2 = -\frac{1}{7}$. Thus $y_1 = -6p + 2$ intersects $y_2 = 8p - 3$ at the point $(\frac{5}{14}, -\frac{1}{7})$. As is clear from Figure 10.8, to maximize the minimum of y_1 and y_2, we take p equal to $\frac{5}{14}$, the p-value of the intersection point. If $p < \frac{5}{14}$, then m, the minimum of y_1 and y_2, is y_2, which is less than $-\frac{1}{7}$; if $p > \frac{5}{14}$, then m, the minimum of y_1 and y_2, is y_1, which is less than $-\frac{1}{7}$. Thus the maximum value of m is $-\frac{1}{7}$, which is attained when $p = \frac{5}{14}$.

To put this result into operation, we must employ a chance device that selects row 1 roughly $\frac{5}{14}$ of the time and row 2 the rest of the time in a random way. One simple chance device can be constructed by putting 5 white marbles and 9 black marbles of the same size into a bag. Reach into the bag and, without looking, pick a marble; if it is white, play row 1 (select heads), and if black, play row 2 (select tails). Over the long run the row player can expect to lose approximately one-seventh of a cent per game, and this is the best he can expect to do.

More generally, to **maximize** *m*, the minimum of the expected values y_1 and y_2 with respect to the column player playing columns 1 and 2, respectively, take for *p* the *p*-value of the point of intersection of y_1 and y_2.

To obtain this value, called the **row player's optimal mixed strategy**, set y_1 equal to y_2, and solve for *p*.

The common value of y_1 and y_2 for this optimal value of *p*, called the *value of the game*, describes how well the row player can expect to do over the long run when rows 1 and 2 are played at random by means of his optimal mixed strategy. Here, too, a game with value zero is called a *fair game*.

The Column Player's Optimal Mixed Strategy

The payoff matrix for the row player is:

$$\begin{bmatrix} -4 & 5 \\ 2 & -3 \end{bmatrix}$$

To obtain the payoff matrix for the column player describing the payoffs of the row player to the column player, we change the sign of the values in this matrix. This yields

$$\begin{bmatrix} 4 & -5 \\ -2 & 3 \end{bmatrix}$$

and the column player's matrix.

Let *r* denote the probability with which column 1 is to be played, and 1 - *r* the probability with which column 2 is to be played. This yields the situation described by Figure 12.8. Expected-value concepts, analogous to the ones defined for the row

	Played with probability r	Played with probability $1 - r$

$$\begin{bmatrix} 4 & -5 \\ -2 & 3 \end{bmatrix}$$

Figure 12.8

player, are defined as follows. The **column player's expected value with respect to row 1 being played by the row player** is the sum of the products of the payoffs in row 1 times the probabilities of these payoffs. For this problem, we have:

$$z_1 = 4r + (-5)(1 - r)$$
$$= 9r - 5, \text{ where } 0 \leq r \leq 1$$

Similarly, the **column player's expected value with respect to row 2 being played by the row player** is the sum of the products of the payoffs in row 2 times the probabilities of these payoffs. This yields:

$$z_2 = (-2)r + 3(1 - r)$$
$$= -5r + 3, \text{ where } 0 \leq r \leq 1$$

The interpretation of these expected values is analogous to that given for their row player counterparts. We can interpret z_1 as approximating the average amount won per game by the column player with respect to row 1 being played by the row player, **assuming** that the game is played a large number of times. Similarly, for z_2 replace row 1 by row 2 in the preceding statement.

Let:

$$n = \text{minimum of } z_1 \text{ and } z_2$$

The **column player's optimal mixed strategy** is defined as the value of r for which n is maximized.

A graphical analysis of z_1 and z_2, similar to the kind given for the row player's expected values, shows that **the column player's optimal mixed strategy is the r-value of the point of intersection of z_1 and z_2.** To obtain this value, set z_1 equal to z_2 and solve for r.

For the problem at hand, we obtain:

$$9r - 5 = -5r + 3$$
$$14r = 8$$
$$r = \frac{4}{7}$$

Another Chapter in the Adventures of Mr. Row and Mr. Column

Mr. Row

Determine the row player's optimal strategy and the value of the game for the game defined by the payoff matrix:

$$\begin{bmatrix} 1 & 3 \\ 4 & 2 \end{bmatrix}$$

First let us note that maximin $= 2$, while minimax $= 3$, so that there is no optimal pure strategy for the row player. We must seek an optimal mixed strategy.

Let p denote the probability with which row 1 is to be played, and $1 - p$ the probability with which row 2 is to be played. We have:

$$\begin{array}{l} \text{probability } p \\ \text{probability } 1-p \end{array} \quad \begin{bmatrix} 1 & 3 \\ 4 & 2 \end{bmatrix}$$

The row player's expected values with respect to the column player playing columns 1 and 2 are given by:

$$y_1 = 1p + 4(1 - p) = -3p + 4$$
$$y_2 = 3p + 2(1 - p) = p + 2$$

Setting y_1 equal to y_2 and solving for p yields:

$$-3p + 4 = p + 2$$
$$-4p = -2$$
$$p = \tfrac{1}{2}$$

Thus $1 - p = \frac{1}{2}$, and the row player's optimal mixed strategy is to play row 1 with probability $\frac{1}{2}$ and row 2 with probability $\frac{1}{2}$; that is, play row 1 about one half the time and row 2 the rest of the time in a random manner. This can be done by using a chance device, such as a bag containing two equal-sized marbles of different colors, black and white, for example. Choose a marble, without looking; if it's white, play row 1, and if it's black, play row 2.

The common value of y_1 and y_2 for $p = \frac{1}{2}$ (obtained by substituting $\frac{1}{2}$ for p in y_1 or y_2) is 2.5; thus the value of the game is 2.5.

Mr. Column

Determine the column player's optimal mixed strategy:

Since the row player's payoff matrix is

$$\begin{bmatrix} 1 & 3 \\ 4 & 2 \end{bmatrix}$$

the column player's payoff matrix is:

$$\begin{bmatrix} -1 & -3 \\ -4 & -2 \end{bmatrix}$$

Let r denote the probability with which column 1 is to be played and $1 - r$ the probability with which column 2 is to be played. The column player's expected values with respect to the row player playing rows 1 and 2 are given by:

$$z_1 = (-1)r + (-3)(1 - r) = 2r - 3$$
$$z_1 = (-4)r + (-2)(1 - r) = 2r - 2$$

Setting z_1 equal to z_2 and solving for r yields:

$$2r - 3 = -2r - 2$$
$$4r = 1$$
$$r = \tfrac{1}{4}$$

Thus $1 - r = \tfrac{3}{4}$, and the column player's optimal mixed strategy is to play column 1 with probability $\tfrac{1}{4}$ and column 2 with probability $\tfrac{3}{4}$. The common value of z_1 and z_2 for $r = \tfrac{1}{4}$ is -2.5, the negative of the value of the game.

By playing his optimal mixed strategy, the column player can expect to do no worse than lose an average of 2.5 units per play over the long run.

12.3   Beware the Assumptions

Assumptions, assumptions, watch the assumptions. William Baumol states this caution on the application of game theory to "real" economic problems.

So far the discussion has dealt only with constant-sum games, i.e., only with games in which the behavior of the players has no effect on their combined payoff. Real economic problems are usually of the nonconstant-sum variety. For example, collusion can normally increase the total profits of a pair of duopolists, and two countries can usually do better by getting together than by declaring war on one another. Unfortunately, the theory is in a far less satisfactory state outside the area of the two-person, constant-sum game. [1; 450].

12.4   Nobel Prize

It is worthy of note that the 1994 Nobel Memorial Prize in Economic Science was awarded to the mathematician John Nash and the economists John Harsanyi and Reinhard Selten. Nash was recognized for his contributions to game theory which provides a mathematical foundation for the economic applications developed by Harsanyi and Selten.

For discussion see, for example, Peter Passel [2], Keith Devlin [3], and David Gale [4].

12.5   Food for Thought Questions

For the games defined by the following payoff matrices, which specify the payoffs from the column player to the row player, determine the optimal strategies for the row and column players and the value.

1. $\begin{bmatrix} 1 & 3 \\ 0 & -2 \end{bmatrix}$ 2. $\begin{bmatrix} 4 & 2 \\ 3 & -3 \end{bmatrix}$ 3. $\begin{bmatrix} 2 & 3 & 4 \\ -1 & 3 & 1 \end{bmatrix}$

4. $\begin{bmatrix} 4 & -1 & 0 \\ 2 & 3 & 1 \\ 3 & 2 & -2 \end{bmatrix}$

5. The Row and Column corporations plan to open fast-food restaurants in a new, still developing shopping center. The shopping center is situated on a rectangular site and various locations are available.

(a)   If, it is **assumed**, both corporations choose center sites or both choose off-center sites, each will capture 50 percent of the business. If, it is **assumed** the Row Corporation chooses a center site and the Column Corporation chooses an off-center site, then the Row Corporation will capture 65 percent of the business (to the Column Corporation's 35 percent); if, it is **assumed**, the Row Corporation chooses an off-center site and the Column Corporation chooses a center site, the Row Corporation will capture 40 percent of the business (to the Column Corporation's 60 percent).

State the pay off matrix that describes the situation, and determine the optimal strategies for both players with respect to the assumptions made. What is the value of the game, and how do you interpret this value.

(b)   If the Row and Column corporations implement their optimal strategies, will the anticipated outcome come to pass? Explain.

For the games defined by the following payoff matrices, which specify the payoffs from the column player to the row player, determine the optimal strategies for the row and column players and the value.

6. $\begin{bmatrix} 1 & -1 \\ -1 & 1 \end{bmatrix}$ 7. $\begin{bmatrix} 6 & 2 \\ 3 & 4 \end{bmatrix}$ 8. $\begin{bmatrix} 4 & 2 \\ 2 & 3 \end{bmatrix}$

9. A beverage company can stress taste or low calorie level in its television advertising. Preliminary studies indicate that advertising which focuses on taste is effective on 50 percent of the viewers not over 40 years old who see the ads, and effective on 30 percent of the over-40 group who see the ads, whereas ads based on calorie level are effective on 25 percent of the not-over-40 group, and effective on 60 percent of the over-40 group.

   Assuming that the same proportion of viewers in the two age groups is exposed to the ads, and viewing the situation as a game with the beverage company and the market as opponents, set up the payoff matrix, and determine the beverage company's optimal strategy and the value of the game.

   (a)   What does implementation of this optimal strategy call for, and how is the value of the game to be interpreted?

   (b)   If the beverage company implements its optimal strategy, will the anticipated outcome come to pass? Explain.

12.6   References of Interest?

1. W. J. Baumol, *Economic Theory and Operations Analysis*, 4th ed. (Englewood Cliffs, N. J.: Prentice-Hall, 1977).

2. P. Passel, "Game Theory Captures a Nobel", *The New York Times*, Oct. 12, 1994.

3. K. Devlin, "Mathematician Awarded Nobel Prize", *Focus*, Dec. 1994.

4. D. Gale, "John Nash and the Nobel Prize", *Focus*, April 1995.

13

Math Modeling in Terms of Matrices

13.1 Preface

In addition to being an interesting system in its own right, matrix algebra (see [1; ch. 5] among many others) has many applications to mathematics itself and to the study of the real world. The one that stands out most, and which I give high priority to introducing to my students, is the input-output matrix model which is the focus of the next section.

13.2 Leontief Input-Output Models

Input-output models for economic systems were pioneered by the economist Wassily Leontief, a recipient of the 1973 Nobel Prize in Economics. In **input-output analysis** an economic system is viewed as a collection of interacting industries in which each industry produces an output that serves as raw materials, or input, for the industries of the system and requires input from the industries of the system. Let a_{ij} denote the amount of input (dollar's worth) of commodity i needed to produce \$1 worth of commodity j; the first subscript refers to input, the second to output. Thus, for example, the equation $a_{21} = 0.20$ asserts that 20¢ worth of commodity 2 is needed to produce \$1 worth of commodity 1. For an n-industry economy, the matrix

$$A = \begin{bmatrix} a_{11} & a_{12} \cdots a_{1n} \\ a_{21} & a_{22} \cdots a_{2n} \\ \vdots & \quad \vdots \\ a_{n1} & a_{n2} \cdots a_{nn} \end{bmatrix},$$

called the **input-coefficient matrix of the system**, specifies the amount of each commodity that is needed to produce $1 worth of each commodity. The entries in the first column, for example, specify the inputs required from each of the n industries to produce $1 worth of the commodity produced by industry 1.

The entries in the first row specify the amount of the commodity provided by industry 1 needed to produce $1 worth of the commodities produced by the n industries of the system.

We also **assume** that there is an **open sector in the economy** (consisting of households, for example) that absorbs a noninput demand for the product of each industry and supplies the primary input, labor. Let $d_1, d_2, \ldots, d_n$ denote the demand of the open sector for the commodities produced by industries $1, 2, \ldots, n$ and let $x_1, x_2, \ldots, x_n$ denote the total output (dollar's worth) of industries $1, 2, \ldots, n$.

The product $a_{ij}x_j$ is (the amount of commodity i needed to produce $1 worth of commodity j) x (total dollar's worth of commodity j produced), and thus expresses the input requirement of industry j for commodity i. For example, if $a_{ij} = 0.20$ (20¢ worth of commodity i is needed to produce $1 worth of commodity j) and $x_j = 5000$ ($5000 worth of commodity j is produced), then $0.20(5000) = \$1000$ worth of commodity i is needed to produce commodity j.

Thus $a_{11}x_1$ is the input requirement of industry 1 for commodity 1, $a_{12}x_2$ is the input requirement of industry 2 for commodity 1, $a_{13}x_3$ is the input requirement of industry 3 for commodity 1, and so on. The sum

$$a_{11}x_1 + a_{12}x_2 + \ldots + a_{1n}x_n + d_1$$

is the sum of the input requirements of the n industries and the open sector for commodity 1. For x_1, the total output of industry 1, to satisfy this demand, we must have:

$$x_1 = a_{11}x_1 + a_{12}x_2 + \ldots + a_{1n}x_n + d_1$$

Similarly, for x_2, the total output of industry 2, to satisfy the demand for commodity 2, we must have:

$$x_2 = a_{21}x_1 + a_{22}x_2 + \ldots + a_{2n}x_n + d_2$$

More generally, for x_n, the total output of industry n, to satisfy the demand for commodity n, we must have:

$$x_n = a_{n1}x_1 + a_{n2}x_2 + \ldots + a_{nn}x_n + d_n$$

Thus the conditions that must be satisfied by output levels $x_1, x_2, \ldots, x_n$, of the n industries in the economy to satisfy the demands of the open sector and the industries themselves are expressed by the following system of n equations:

$$x_1 = a_{11}x_1 + a_{12}x_2 + \ldots + a_{1n}x_n + d_1$$
$$x_2 = a_{21}x_1 + a_{22}x_2 + \ldots + a_{2n}x_n + d_2$$
$$\cdot \quad \cdot$$
$$\cdot \quad \cdot$$
$$\cdot \quad \cdot$$
$$x_n = a_{n1}x_1 + a_{n2}x_2 + \ldots + a_{nn}x_n + d_n$$

Rewriting this system so that terms involving $x_1, x_2, \ldots, x_n$ appear on one side and the constants $d_1, d_2, \ldots, d_n$ appear on the other side yields:

$$
\begin{aligned}
(1 - a_{11})\,x_1 - \quad & a_{12}\,x_2 - \ldots - \quad a_{1n}\,x_n = d_1 \\
- a_{21}\,x_1 + (1 - a_{22})\,x_2 - \ldots - \quad a_{2n}\,x_n = d_2 \\
\cdot \qquad\qquad & \cdot \\
\cdot \qquad\qquad & \cdot \qquad\qquad\qquad (1) \\
\cdot \qquad\qquad & \cdot \\
- a_{n1}x_1 \qquad\quad & - a_{n2}x_2 - \ldots + (1 - a_{mn})k_n = d_n
\end{aligned}
$$

It is advantageous, as we shall see, to express this system in terms of a matrix equation involving a matrix product. To do so we introduce matrices, I_n, X, and D, as follows:

$$I_n = \begin{bmatrix} 1 & 0 & 0 & \cdots & 0 \\ 0 & 1 & 0 & \cdots & 0 \\ 0 & 0 & 1 & \cdots & 0 \\ \vdots & \vdots & \vdots & & \vdots \\ 0 & 0 & 0 & \cdots & 1 \end{bmatrix}, \quad X = \begin{bmatrix} x_1 \\ x_2 \\ \vdots \\ x_n \end{bmatrix}, \quad D = \begin{bmatrix} d_1 \\ d_2 \\ \vdots \\ d_n \end{bmatrix}$$

Matrix I_n is an n by n matrix with 1's in the main diagonal and 0's elsewhere; X is called the **output matrix** of the system and D is called the **final-demand matrix** of the system.

Subtracting A, the input-coefficient matrix, from I_n yields:

$$I_n - A = \begin{bmatrix} (1-a_{11}) & -a_{12} & \cdots & -a_{1n} \\ -a_{21} & (1-a_{22}) & \cdots & -a_{2n} \\ \vdots & & \vdots & \\ -a_{n1} & -a_{n2} & \cdots & (1-a_{nn}) \end{bmatrix}$$

Taking the product of $(I_n - A)$ and X, $(I_n - A)X$, yields the left side of system (1); the right side of system (1) is expressed by matrix D. Thus, in matrix terms, system (1) is expressed by the following matrix equation:

$$(I_n - A)X = D$$

In summary, then, the problem of satisfying the needs of the n industries of the economy, expressed by the input-coefficient matrix A, and the needs of the open sector of the system, expressed by the final-demand matrix D, reduces to the matrix problem of determining output matrix X such that the product $(I_n - A)X$ equals D.

> As long as the input-coefficient matrix A does not change, $(I_n - A)^{-1}$ does not change, and with one matrix inversion a variety of possible final-demand situations can be studied.

To illustrate, consider a two-industry economy governed by the input-coefficient matrix

$$A = \begin{bmatrix} 0.4 & 0.3 \\ 0.3 & 0.2 \end{bmatrix}$$

and having final-demand matrix $D = \begin{bmatrix} d_1 \\ d_2 \end{bmatrix}$. The problem is to find an output matrix $X = \begin{bmatrix} x_1 \\ x_2 \end{bmatrix}$ such that $(I_n - A)X = D$. If $(I_2 - A)^{-1}$ exists, then $X = (I_2 - A)^{-1} D$.

$$I_2 - A = \begin{bmatrix} 0.6 & -0.3 \\ -0.3 & 0.8 \end{bmatrix}, \quad (I_2 - A)^{-1} = \begin{bmatrix} \dfrac{80}{39} & \dfrac{10}{13} \\ \dfrac{10}{13} & \dfrac{20}{13} \end{bmatrix}$$

Thus

$$X = (I^2 - A)^{-1}D = \begin{bmatrix} \dfrac{80}{39} & \dfrac{10}{13} \\ \dfrac{10}{13} & \dfrac{20}{13} \end{bmatrix}\begin{bmatrix} d_1 \\ d_2 \end{bmatrix} = \begin{bmatrix} \dfrac{80}{39}d_1 + \dfrac{10}{13}d_2 \\ \dfrac{10}{13}d_1 + \dfrac{20}{13}d2 \end{bmatrix}$$

and we have:

$$x_1 = \frac{80}{39}d_1 + \frac{10}{13}d_2$$

$$x_2 = \frac{10}{13}d_1 + \frac{20}{13}d_2$$

If $d_1 = \$39,000$ and $d_2 = \$61,000$ ($\$39,000$ worth of commodity 1 and $\$61,100$ worth of commodity 2 are required by the open sector), then $x_1 = \$127,000$ and $x_2 = \$124,000$. $\$127,000$ worth of commodity 1 and $\$124,000$ worth of commodity 2 must be produced to satisfy the input needs of industries 1 and 2 and the requirements of the open sector.

If the projected requirements of the open sector should change to $d_1 = \$46,800$ and $d_2 = \$65,000$, then $x_1 = \$146,000$ and $x_2 = \$136,000$. $\$146,000$ worth of commodity 1 and $\$136,000$ worth of commodity 2 must be produced to satisfy the input needs of industries 1 and 2 and the requirements of the open sector.

The successful application of the input-output model to an economy requires that a **realistic input-coefficient matrix A and final demand matrix D** be developed for the economy.

 Assuming that this could be done, **the big IF**, the second major problem is to suitably refine these matrices so that they are **realistic over time.**

13.3   Never Lose Sight of the Assumptions

The afore **assumption** is not one that should be taken lightly. Hollis Chenery and Paul Clark [3] and William Baumol [2] note:

The properties of Leontief models can be derived from **three basic assumptions**, which it will be useful to state at the outset:

---

(1) *Each commodity (or group of commodities) is supplied by a single industry or sector of production.* Corollaries of this assumption are (*a*) that only one method is used for producing each group of commodities; and (*b*) that each sector has only a single primary output.

(2) *The inputs purchased by each sector are a function only of the level of output of that sector.* (The stronger assumption is usually made that the input function is linear, but this is a matter of convenience.)

(3) *The total effect of carrying on several types of production is the sum of the separate effects.* This is known as the additivity assumption, which rules out external economies and diseconomies.

The validity [meaning **realism**, here] of each of these **assumptions** depends both on the nature of production in single plants and on the way in which these units are aggregated into sectors. Some **assumptions** may be more valid [that is, **realistic**] for aggregates than for individual units, as for example the exclusion of joint products and external economies. Others may hold for single productive processes but not for sectors. We must therefore consider together the nature of the underlying production relationships and the effects of aggregation in evaluating the structure of the model. [3; 33-4]

Perhaps more serious is a second assumption which states that in any productive process all inputs are employed in rigidly fixed proportions and the use of these inputs expands in proportion with the level of output. This is a special case of an assumption of constant returns to scale (see Chapter 11, Section 5). But the fixed-proportions assumption is far more restrictive. Constant returns to scale is perfectly consistent with the substitution of one factor for another. A linear homogeneous production function (constant return to scale) permits both labor-intensive and capital-intensive processes. The firm whose production function exhibits constant returns can if it wishes have one hundred workers for every $1,000 invested in machinery, or it may use machines which require only ten workers per $1,000 machine investment. A linear homogeneous production function requires only that if the firm decides to triple the scale of either of these types of operation, the result will be a tripling of output. Not so the Leontief fixed-proportions premise, which requires that a manufacturing process which is labor intensive offer no option of a capital-intensive alternative.[3] If fifty-three men per $1,000 of investment are required at any level of operation, it is assumed that the same ratio will be required no matter how much the size of the firm expands or contracts. Whether this assumption is relatively innocuous or does considerable violence to the input-output results is still under dispute. But the premise is certainly never absolutely true, even in those cases where chemistry and engineering dictate fixed proportions between some ingredient and output. [2, 538-9.]

13.4 Food for Thought Questions

1. For a two-industry economy, let us assume that $0.20 and $0.30 worth of the first industry's commodity is needed by the first and second industries to produce $1 worth of their respective commodities and that $0.30 and $0.10 worth of the second industry's commodity is needed by the first and second industries to produce $1 worth of their respective commodities.

    (a)   Set up the input-coefficient matrix of the economy.

    (b)   If the open sector of the economy requires $2520 worth of commodity 1 and $3150 worth of commodity 2, what output levels will satisfy the input needs of the industries and the requirements of the open sector?
          How much of these outputs will be consumed by industries 1 and 2?

    (c)   If the open sector of the economy requires $3465 worth of commodity 1 and $3780 worth of commodity 2, what output levels will satisfy the input needs of the industries and the requirements of the open sector?
          How much of these outputs will be consumed by industries 1 and 2?

Question 2 is another version of 1; the numbers have been changed.

2. For a two-industry economy, let us assume that 50¢ and 20¢ worth of the first industry's product is needed by the first and second industries to produce $1 worth of their respective commodities, and that 20¢ and 50¢ worth of the second industry's product is needed by the first and second industries to produce $1 worth of their respective commodities.

    (a)   Set up the input-coefficient matrix of the economy.

    (b)   If the open sector of the economy requires $1575 worth of commodity 1 and $1890 worth of commodity 2, what output levels will satisfy the input needs of the industries and the requirements of the open sector?

How much of these outputs will be consumed by industries 1 and 2?

(c)   If the open sector of the economy requires $2940 worth of commodity 1 and $3675 worth of commodity 2, what output levels will satisfy the input needs of the industries and the requirements of the open sector?

How much of these outputs will be consumed by industries 1 and 2?

13.5   References of Interest?

1.   W. J. Adams, *Finite Mathematics, Models, and Structure*, Revised Edition (Philadelphia: Xlibris, 2009); available on the web at webpage.pace.edu/wadams.

2.   W. J. Baumol. *Economic Theory and Operations Analysis*, 4th ed. Englewood Cliffs, N.J.: Prentice-Hall, Inc., 1977, Chapter 22.

3.   H. B. Chenery and P. G. Clark. *Interindustry Economics*. New York: John Wiley & Sons, Inc., 1959.

4.   Conference on Research in Income and Wealth, National Bureau of Economic Research. *Input-Output Analysis: An Appraisal*. Princeton, N.J.: Princeton University Press, 1955.

5.   R. Dorfman, P. A. Samuelson, and R. M. Solow. *Linear Programming and Economic Analysis*. New York: McGraw-Hill Book Company, 1958, Chapters 9-12.

6.   W. W. Leontief. *The Structure of the American Economy, 1919-1939*, 2d ed. New York: Oxford University Press, 1951.

7.   W. W. Leontief, ed. *Studies in the Structure of the American Economy*. New York: Oxford University Press, 1953.

14

Tales from the Land of Differential Equation Modeling

14.1   Preface

In many situations of interest a mathematical model is formulated based on the **assumption** that the instantaneous rate of change $\frac{dy}{dt}$ of the amount y of a quantity with respect to time t is proportional to the amount of y that is present at time t. In differential equation terms, this translates to

$$\frac{dy}{dt} = ky$$

or, equivalently,

$$f'(t) = kf(t)$$

where k is the constant of proportionality, and $y = f(t)$ is the function to be determined. We shall solve this equation and then discuss some applications.

Dividing both sides of $f'(t) = kf(t)$ by $f(t)$ yields:

$$\frac{f'(t)}{f(t)} = k$$

Thus:

$$\int \frac{f'(t)}{f(t)}\, dt = \int k\, dt$$
$$\ln f(t) = kt + C \tag{1}$$

Let N_0 denote the amount of y that is present when $t = 0$ (the beginning of the process). Substituting N_0 for $f(t)$ and 0 for t in (1) yields:

$$\ln N_0 = C$$

Replacing C by $\ln N_0$ in (1) gives us:

$$\ln f(t) = kt + \ln N_0$$

Subtracting $\ln N_0$ from both sides yields:

$$\ln f(t) - \ln N_0 = kt$$

Since the logarithm of a quotient is the difference of the logarithms of the components, we can write the above as follows:

$$\ln \frac{f(t)}{N_0} = kt$$

To say that $\log_a M = w$ means, by definition, that $a^w = M$. Using this here with $M = \dfrac{f(t)}{N_0}$, $w = kt$, and $a = e$ gives us

$$e^{kt} = \frac{f(t)}{N_0}$$

or

$$f(t) = N_0 e^{kt} \tag{2}$$

where k is the constant of proportionality (determined from data intrinsic to the nature of the process) and $N_0 = f(0)$, the value of the function at the beginning of the process ($t = 0$).

An interesting applications of this model type is to radiocarbon dating.

14.2   Radiocarbon Dating

Earth is continually bombarded by sub-atomic particles, termed cosmic radiation, which are emitted by the Sun. These highly energetic particles react with atoms in the atmosphere to produce neutrons, which then collide with nitrogen atoms in the atmosphere to produce a radioactive form of carbon called carbon-14 or radiocarbon. Carbon-14 decays spontaneously, giving off an electron, and changing to nitrogen. This decay process is such that it takes about 5,730 years for half the amount of carbon-14 in a substance to disintegrate. This value is called the **half-life of carbon-14**. Carbon-14 behaves chemically in the same way as ordinary carbon-12.

Living plants and animals absorb carbon-14 along with carbon-12. When they die this absorption ceases, but the radioactive disintegration of carbon-14 continues, slowly at a fixed known rate. The simple idea that emerges, at least in theory, is based on measuring the proportion of carbon-14 left in a sample whose age is to be determined. Since the initial proportion of carbon-14 in the substance is known, to a good approximation, when it was living, we should be able to calculate how long the radioactive decay process had been going on, which gives us the age of the material.

Shortly after World War II Willard F. Libby proposed a way of employing the radioactive decay of carbon-14 to estimate the age of substances containing carbon-14, particularly organic remains. This includes charcoal, wood, cloth, limestone, bones, hair and soil. Libby's carbon-14 dating technique, for which he was awarded the 1960 Nobel Prize in chemistry, is based on the following **assumptions**:

1. Carbon-14 is produced in the atmosphere at an approximately constant rate.

2. Except for recent geologic time (the last half-century), there is a constant concentration of carbon-14 in all living things, which is about 1 carbon-14 atom to 1 trillion carbon-12 atoms.

3. The rate at which carbon-14 decreases is proportional to the amount present.

These **assumptions** lead to an exponential decay model

$$\frac{dy}{dt} = ky,$$

where $y = f(t)$ expresses the weight (in suitable units) of carbon-14 in a given portion of matter and t is time measured in years.

If we let $N_0 = f(0)$ denote the amount of carbon-14 present at time $t = 0$ (the beginning of the process), then from (2) we obtain the function

$$y = N_0 e^{kt}. \tag{3}$$

If the value of k is known, then $y = N_0 e^{kt}$ enables us to compute the value of y for each given t, or the value of t for each $y > 0$. The value of k can be obtained by using the fact that the half-life of carbon-14 is 5730 years; that is, the length of time it takes half of the carbon-14 nuclei in a sample to decay is 5730 years. In other words, when $t = 5730$, $y = N_0/2$. Substitution into (3) gives us:

$$\frac{N_0}{2} = N_0 e^{5730k}$$

$$\frac{1}{2} = e^{5730k}$$

In logarithm form this yields:

$$5730k = ln(1/2) \simeq -0.69315$$
$$k \simeq -0.0001209$$

Replacing k by -0.0001209 in (3) yields

$$y = N_0 e^{-0.0001209t} \qquad (4)$$

as the function describing the amount of carbon-14 in a substance in terms of time. The graph of (4) is shown in Figure 14.1.

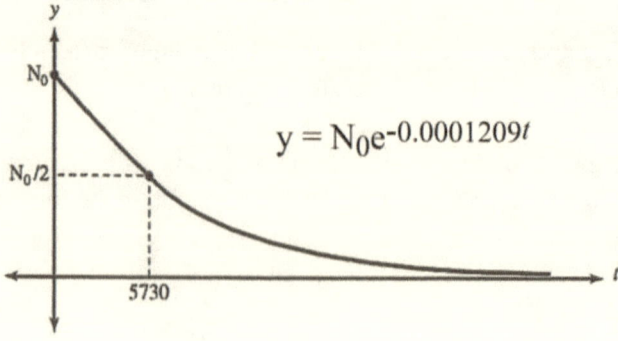

Figure 14.1

This is what the model says, but what does reality say? Libby put his method to the test in 1955 when he obtained radiocarbon dates of a number of samples, mostly from Egypt, whose ages had been determined by other means. The radiocarbon dates were close to the established dates, and a powerful new dating tool was made available to archeologists. This tool made possible decisive advances. Colin Renfrew notes that one of the greatest contributions of the first radiocarbon revolution was in making possible the study of world prehistory. Developments throughout the world may now be studied on a comparative basis with a sound framework of dates. [5]

Age of a Mummy

The skin, bone, and clothing of an adult mummy found in a cave near Lake Winnemucca, Nevada, was found to contain 74 percent of the original carbon-14. What is the age of the mummy?

Substituting $0.74 N_0$ for y in (4) yields:

$$0.74 N_0 = N_0 e^{-0.0001209t}$$

$$0.74 = e^{-0.0001209t}$$

By definition of natural logarithm, this means:

$$\ln 0.74 = -0.0001209t$$

$$t = \frac{\ln 0.74}{-0.0001209}$$

$$\simeq -\frac{0.3010}{-0.0001209}$$

$$= 2490$$

Thus the mummy is approximately 2500 years old.

For a proper perspective on this (valid) conclusion, it is important for us to keep in mind the afore **three assumptions** on which this conclusion is founded. These **assumptions** are considered in sections 14.3 and 14.4.

14.3   Model vs. Reality: 1

The basic radiocarbon dating model has undergone a number of refinements. Libby's **second assumption**, for one, is open to question. The amount of carbon-14 in the atmosphere has not been constant with time. It has varied by as much as ±5% because of changes in solar activity and Earth's magnetic field. In recent years contamination from the burning of fossil fuels and testing of nuclear weapons has resulted in significant changes in the amount of carbon-14 in the atmosphere. Studies of the bristlecone pine, a tree that grows in the White Mountains of California and lives for up to 5000 years, has allowed scientists to develop calibration curves for carbon-14 dates to correct for changes in the level of carbon-14 over time. Accurate tree ring records of age are available for a period as far back as 9000 years and scientists have sought other indicators of age against which carbon-14 dates can be compared.

One such indicator is a uranium-thorium-230. Recently conducted studies at the Lamont-Doherty Geological Laboratory of Columbia University indicate that age estimates using carbon dating and uranium-thorium dating were in basic agreement for the period from 9000 years ago to the present. For earlier times the carbon-14 dates were substantially younger than those obtained by uranium-thorium analysis. The largest deviation, 3500 years, was obtained for samples that were approximately 20,000 years old. These results make clear some of the limitations of carbon dating and may lead to revisions in the age estimates that have been assigned by carbon dating to a number of "older" samples. [2]

14.4   Model vs. Reality: 2

As successful as the Libby Model has been in helping us to get a grip on world prehistory, it has the startling feature of being in contradiction to the basic nature of the radioactive decay process—a continuous model in contract to a discrete process.

The weight y of the carbon-14 sample suddenly decreases when a particle is radiated and otherwise remains constant. This feature leads to a graph of $y = N(t)$ of the sort shown in Figure 14.2 which

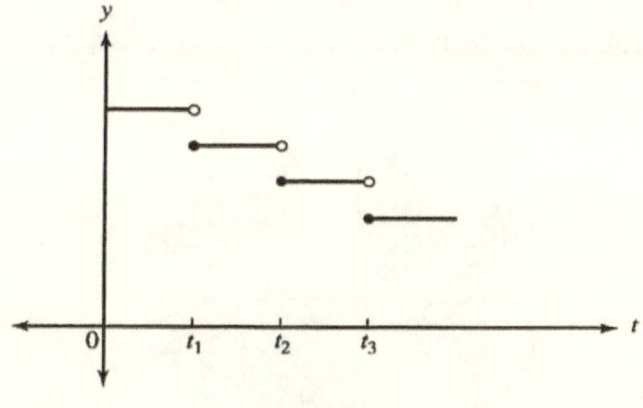

Figure 14.2

is that of a step function discontinuous at those instances t_1, t_2, . . . at which a particle is radiated, and otherwise constant.

In a sense this fact proves the first description wrong, but it would be appropriate to view the new model as a suitable replacement for the exponential decay model when we are concerned with far greater detail than the exponential decay model could provide us with. The function (4), suitably refined, is accurate for macroscopic predictions, that is, for predictions involving time intervals which go back as far as 10,000 years; its accuracy falls of for time periods exceeding 10,000 years and gives poor results in connection with microscopic predictions involving a "small number" of atoms of carbon-14.

14.5   Radioactive Decay Vindicates Van Meegeren

An interesting application of radioactive decay modeling is provided by the problem of detecting art forgery.

On May 29, 1945 the Dutch painter H.A. Van Meegeren was arrested on a charge of collaboration with the Nazis in having sold the priceless Vermeer painting "Christ and the Adulteress" to Herman Goering. Van Meegeren claimed that he was not guilty of aiding the enemy in acquiring priceless Dutch art because he himself had painted "Christ and the Adulteress," the famous "Supper at Emmaus," as well as other paintings attributed to Vermeer and the less famous de Hooghs. The art world was understandably skeptical, and to prove his point Van Meegeren began, while in prison, to paint the

Vermeer painting "Jesus Amongst the Doctors". When the work was nearly completed the charge of collaboration was changed to the less serious charge of forgery. Van Meegeren then refused to finish and age the painting in the

hope of thwarting the investigation. An international panel of experts that was appointed to investigate the matter concluded that the alleged Vermeers were forgeries. Van Meegeren was vindicated and was sentenced to a year in prison for forgery. While in prison he died of a heart attack on December 30, 1947. In spite of the evidence gathered by the panel of experts, many people refused to believe that "Supper at Emmaus" was a Vermeer forgery. It did not seem possible that such a masterpiece could have been forged by a painter considered third rate. More conclusive proof was demanded, and in 1967 scientists at Carnegie Mellon University took up the problem.

Their analysis was based on the radioactive decay of lead-210 and radium-226, small amounts of which are found in the widely used pigment lead oxide. Lead-210 has a comparatively short half-life of 22 years and if the paint used in painting is very old in comparison to this short half-life, the amount of radioactivity from the lead-210 in the paint will be approximately equal to the amount of radioactivity due to the radium; if the painting is recent, the amount of radioactivity from the lead-210 will be much greater than that generated by the radium. The Carnegie Mellon group made this precise by employing a mathematical model similar to, but more complex than, the one employed for carbon-14 decay. They showed that the amount of radioactivity from lead-210 is much greater than that from radium-226.

In the end van Meegeren was in a sense vindicated; he was not a Nazi collaborator and he did paint a work that had been attributed to Vermeer; [3], [4], and [6].

For discussion of exponential growth modeling I recommend [1; ch. 11].

14.6   Food for Thought Questions

1. Artifacts obtained in an unearthed settlement are found to contain 60 percent of the original amount of carbon-14. Find the date of the artifacts. Note: ln 0.6 ≃ -0.5108. What **assumptions** should we keep in mind?

2. Radium is an unstable element which undergoes radioactive decay. Let $y = N(t)$ denote the number of grams of radium in a given portion of matter in terms of time t, in years. Set up a differential equation model for $y = N(t)$ based on the **assumption** that the rate at which radium decreases is proportional to the amount present. Taking 1,600 years as the half-life of radium, determine $y = N(t)$. What **assumptions** should we keep in mind?

3. Bones found in a tomb are found to contain 80% of the original amount of carbon-14. Based on carbon-14 dating, approximately how old are the bones? Note: ln 0.8 ≃ -0.22314. What **assumptions** should we keep in mind?

14.7   References of Interest?

1. W. J. Adams, *Fundamentals of Calculus with Applications and Companion to Calculus* (Philadelphia: Xlibris, 2008); available on the web at webpage.pace.edu/wadams.

2. M. Browne, "Errors are Feared in Carbon Dating," *The New York Times*, May 31, 1990.

3. P. Coremans, *Van Meegeren's Faked Vermeers and DeHooghs* (Amsterdam: Meulenhoff, 1949).

4. E. Dolnick, *The Forger's Spell: A True Story of Vermeer, Nazis, and the Greatest Art Hoax of the Twentieth Century* (New York: Harper, 2007)

5. C. Renfrew, *Before Civilization: The Radiocarbon Revolution and Prehistoric Europe* (New York: Knopf, 1973).

6. M. Simons, "A Most Artful Forger Now Beguiles the Dutch," *The New York Times*, May 18, 1996; 4.

15

Math Modeling for the Study of the Motions of Celestial Bodies

15.1 Preface

In the late 1600s a great argument took hold in London's scientific circles over what principle would account for the elliptical orbits of the planets. The astronomer Edmund Halley put the question to Isaac Newton who immediately replied, "the inverse square principle." That is, what is today called Newton's principle of universal gravitation: Every object in the universe attracts every other object with a force which varies directly as the product of their masses and inversely as the square of the distance between them. "How do you know?" asked Halley. "I calculated these orbits from this principle," replied Newton.

Newton presented these principles in his treatise, *Mathematical Principles of Natural Philosophy* (1687) [1]. In his *Principia*, as it is referred to from its Latin title, Newton brilliantly synthesized the work of Copernicus, Kepler, Galileo and others, adding to it the product of his own genius, to formulate what we today term a mathematical model—Newtonian Mechanics—to describe the behavior of terrestrial and celestial objects in motion.

Newton's Other "Laws" (Postulates) of Motion

The following observations might be of interest to my colleagues teaching the afore material.

The question I should like to comment on is: Should Newton's other postulates ("Laws") of motion be explicitly stated in class discussion of the afore topic?

I would have to say that it depends on circumstances. If your audience has not studied basic physics (as was the case with classes I taught) there is nothing positive to be gained by rushing through Newton's postulates. From an education point of view I believe it would be counterproductive to do so. To obtain a grip on Newton's postulates requires a fair amount of time, which would serve as a distraction from the main point to be made about math modeling. I point out that Newton's postulates is a group of postulates that underlie what is termed Newtonian mechanics, suggest that we take it from this point, consider some of its predictions (theorems) and the question of their realism.

If your audience has studied basic physics, then stating Newton's postulates to refresh memories would not be amiss.

15.2   Ceres: A Triumph of Newtonian Mechanics

A New Planet Has Been Discovered, But Where is It?

In the summer of 1801 the discovery of a new planet was excitedly reported by the press. This new planet, called Ceres, was first sighted on New Years Day by Giuseppi Piazzi at the Palermo Observatory. He was able to observe it only until February 11th, after which it was hidden by the Sun. Piazzi's attempts to compute the orbit of Ceres from the few position readings he had managed to obtain were unsuccessful.

Attempts were made by many European astronomers to determine its orbit and predict the location of the new planet as it emerged from the Sun's cover, but without success. It seemed that Ceres had been lost.

An On-target Prediction

Carl Friedrich Gauss (1777-1855), called the Prince of Mathematicians, was at the beginning of a long and distinguished career in mathematics and science at this time. Gauss, who had already achieved recognition for his work in number theory, took up the problem of determining the orbit of Ceres in November of 1801. He communicated his conclusions, obtained from **Newton's postulates** (generally called Laws of Motion), to the astronomers Franz von Zach and Heinrich Olbers who located Ceres in the closing days of 1801 where Gauss' deductions predicted it would be found. This chain of developments added further support to Newton's theory of planetary motions.

Further observations of Ceres by Olbers led to his discovery of another planet, Pallas, in April of 1802 in the vicinity of Ceres. Gauss sustained his reputation as a mathematical astronomer by determining the orbit of Pallas. In 1804 a third planet, Juno, was discovered by Ludwig Harding and, as the recognized expert in such matters, the calculation of its orbit fell to Gauss as well. Ceres, Pallas, and Juno were the first discovered members of the asteroid belt, a numerous group of very small planets, called planetoids or asteroids, revolving about the Sun in the region between Mars and Jupiter.

15.3   Further Challenges and Triumphs

Trouble in the Heavens: Uranus and Neptune

By 1840 six major planets of the Solar System were known, Uranus, discovered in 1781, being the latest to be found. The orbit of Uranus was determined from widely separated observations of its position and the expected deviations from its elliptical path due to the pull of the other planets were accurately predicted by Newton's principles. Uranus takes eighty-four years to complete a revolution of its orbit and for fifty years after having been discovered the planet behaved according to predictions. By 1830 significant differences between the actual behavior of Uranus and the path predicted for it by Newtonian mechanics had become evident. Reality and the mathematical model predicting its nature were in disagreement and with such confrontations the model always loses; it must back down. The question is, in what way?

Newtonian Mechanics Triumphant

There was the possibility, believed by some, that Newton's principles were not applicable over such large distances. Another possibility, which had its supporters, was that a hitherto undiscovered planet was exerting a gravitational pull on Uranus and causing the noted deviation from its predicted orbit. This possibility was independently taken up by John Couch Adams (1819-1892), while still an undergraduate at St. John's College of Cambridge University, and the French mathematical astronomer Urbain Leverrier (1811-1877). Adams solved the difficult mathematical problem of locating the position of this suspected planet from the observed motion of Uranus and Newtonian principles by September 1845. He determined the elements of the orbit of the suspected planet as well as its mass and position for 1 October 1845. Unfortunately, curious circumstances which fall under the framework of Parkinson's principle (whatever could go wrong will go wrong) led to misunderstanding and Adams was unsuccessful in having the Greenwich Observatory search the heavens for the suspected planet where his deductions predicted it to be. If it had been searched for then, it would have been found very close to where Adams said it would be found.

Leverrier, who had an established reputation in the astronomical world, took up the problem and published his conclusions in a series of papers communicated between November 1845 and August 1846. The planet, subsequently named Neptune, was found by the Berlin Observatory on 23 September 1846, almost a year after Adams' prediction of its location. Adams was the first to obtain a mathematical deduction of Neptune's position, but Leverrier was the first to have his deduction verified by observation. Today the brilliant independent work of both mathematical astronomers is fully recognized. At the time a bitter dispute broke out over who had priority, illustrating again that the affairs of the planets, complex as they may be, are less tempestuous than those of mortals on Earth.

It was an extraordinary display of the power of mathematics—no data on the planet, but it must be there and there it is—which lent further support to Newton's mathematical model of planetary motions as a realistic model for the behavior of celestial bodies.

Another Test of Realism

The orbit of Neptune was determined and the planet, which takes 164.8 years to complete a revolution in its orbit, was closely observed. In time the deviations of this planet and Uranus as well became larger than expected on the basis of known forces and the Newtonian model. An undiscovered planet was the cause of this behavior before and it seemed "reasonable" that a hitherto undiscovered planet was the cause in this situation as well. Percival Lowell (1855-1916) carried out the difficult deductions and after an arduous search the predicted planet, named Pluto, was found by the observatory in Arizona that he had founded.

Thus another great triumph was scored by Newtonian mechanics, but trouble was brewing in the heavens for the Newtonian model in the behavior of Mercury and light.

15.4   Mercury and Light: Troublemakers

According to valid consequences of Newton's model for the planetary motions
a planet revolves around the Sun in a stationary elliptical orbit, with other
planets producing perturbations which could be calculated and observed.

It has been long known that the orbit of Mercury does not remain stationary
as predicted by Newton's model. The path of Mercury is very nearly an ellipse,
but it does not close up in one revolution. In the next revolution the path has
advanced slightly in the same direction in which the planet was moving. The
orbit is an ellipse which is itself slowly revolving as shown in Figure 15.1.

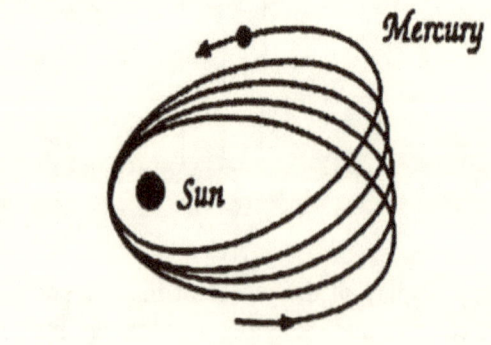

Figure 15.1

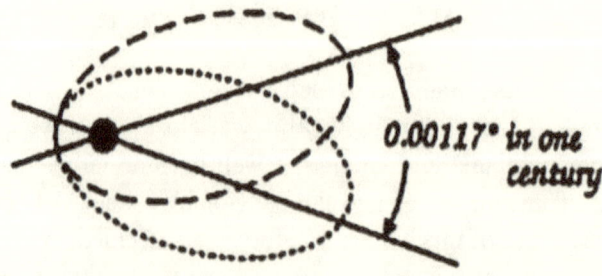

Figure 15.2

The amount of the shift is so small in terms of a year that it could not be detected, but if we let it add up for a century it comes to 0.00117 degrees per century, as indicated by Figure 15.2

How is the discrepancy between the observed shift in elliptical path and the prediction of Newton's model to be accounted for? Leverrier thought that the peculiar behavior of Mercury might be due to an interior planet which he tentatively called Vulcan. But, although thoroughly searched for, Vulcan was never found and Leverrier's hypothesis was never confirmed.

A ray of light bends when it passes near a "heavy" mass, such as the Sun. The amount of the deflection predicted by Newtonian Mechanics is 0.875 arcseconds (3600 arcseconds = 1^0). The determination of reality's verdict requires rather special conditions—a total eclipse of a sun at a point where a large number of bright stars are in the background. Such an eclipse took place on 19 May 1919; the data obtained yielded a deflection value of 1.98 ± 0.12 arcseconds [2; ch. 4], which establishes that Newton's value is considerably off reality's mark.

15.5   Einstein's Model Triumphant

In 1916 Albert Einstein (1879-1955) published his General Theory of Relativity, a mathematical model which has as one of its theorems the deduction that the orbit of Mercury should shift by 0.00119 degrees per century. [2; ch. 5] Einstein's result is close to the 0.00117 degree observed value, and his model entered the scene as a possible refinement of Newton's model.

Another clear point of disagreement between the Newtonian and Einsteinian models concerns the behavior of light.

> Newtonian model: predicted deflection = 0.875 arcseconds
> Einsteinian model: predicted deflection = 1.75 arcseconds,

which is twice as large as the Newtonian model's value.

Reality's mark = 1.98 ± 0.012 arcseconds which, comparatively speaking, favors Einstein's model (13% deviation) over Newton's (26% deviation).

Einstein's Postulates

If you were to put a gun to my head and demand that I state Einstein's postulates or . . . , I would have to hope that I'm not too badly off with the alternative.

As with Newton's postulates, lacking the explicit statement of Einstein's postulates does not deter us from talking about Newton's model vs. Einstein's model and how they compare as descriptions of reality.

15.6   Perspective on Math Models

Newton's model met the challenges hurled at it by reality for over two hundred years. One troublemaker was Mercury, whose orbit was found not to agree with the predictions of Newton's model; another was light. Attempts to reconcile the behavior of Mercury and light with Newton's model were unsuccessful. His model was succeeded by Einstein's model, his General Theory of Relativity, which accurately predicted the behavior of nature in other circumstances as well as Newton's model. Einstein's achievement in refining Newton's model, we should observe, stands as one of the great landmarks of human thought. Nevertheless, should Einstein's model find itself at odds with the behavior of nature it too will have to be refined.

The process of refining mathematical models for natural world and institutional phenomena is a central feature of the mathematical modeling process. Sir Rudolph Peierls offers the following insightful analogy to help illustrate the evolutionary nature of model building.

> If we look at the photograph of a landscape in snow, showing part of a snowfield in sunlight and another part in the shade, we would describe this photograph as having one area in very light shade of grey, almost white, and another uniform area of a slightly darker shade of grey. On a more careful study of the same photograph under a microscope we discover that the grey areas are in fact made up of small black dots, the silver grains of the photographic process, on a white background, the darker area differing from the lighter by having more of the black dots.
>
> In a sense this discovery has proved the first description wrong, but it would be more reasonable to say that the new description refines the old one and replaces it when we are concerned with far greater detail than was the case at first. The old description is still good enough when we are concerned with taking a photograph or with looking at it. In fact by being simpler it is more valuable for that purpose. If we had to think of photographs always as collections of black dots on a white background we should find the photograph quite useless as pictures [1; p. 17].

15.7   Food for Thought Questions

1. Consider Rasa's experience vs. Andy's model ("Rasa's Vacation Trip," Ch. 2) and the behavior of Mercury and light vs. Newton's model.
    Are there similarities in the two situations that are noteworthy? Explain.

2. What was the aftermath of Rasa's experience and the observed behavior of Mercury and light? What similarity, if any, could you point to? Explain.

3. Einstein is reported to have commented, "No amount of experimentation can ever prove me *right*; a single experiment may at any time prove me *wrong*."
    Prove me *right*, prove me *wrong*; in what sense? What did he mean?

4. "A refined radar technique that may settle the current debate over the validity of Einstein's general theory of relativity has been successfully tested." (*The New York Times*, Feb. 28, 1968, p. 20.)

    (a)   Could such a technique be used to settle a question of validity?

    (b)   What issue might such a technique help to resolve?

5. "The most accurate long-distance measurements ever made, by means of radio signals between the Viking spacecraft on Mars and antennas on Earth, have produced new confirmation of Einstein's theory of relativity, a Viking project scientist reported today." (*The New York Times*, Jan. 7, 1977, p. A8.) In what sense was the theory of relativity confirmed?

The views stated in 6 were the replies of two students to the question stated in 6. I thought it would be instructive to submit them to you for your consideration.

6. The question is, how did the behavior of Mercury send a signal that Newton's model for the behavior of the planets had to be modified?
    Consider the following two replies to this question. State whether you Agree or Disagree, and give the basis for your conclusion. If you disagree with both, how would you answer the question?

    Newton's own postulates on the behavior of Mercury and light were proved to be unrealistic by Einstein's assumptions. Newton assumed Mercury travels in a stationary elliptical orbit, but Einstein later deduced that Mercury's orbit shifts slightly, 0.00119 degrees per century. Newton made the assumption that light is bent by large massive bodies by a certain number of degrees. But Einstein proposed that the number of degrees light is bent is twice that assumed by Newton.

    Mercury and light were troublemakers for Newton's model of planetary motions because their behavior showed his conclusions not to be valid.

7. What would you say in response to the following view? Newton devised a theory of planetary motions which scientists believed was true for over

two hundred years. How could we have confidence in the theories of science when a theory believed in for so long turned out to be false?

8. One valid consequence of Einstein's relativity theory is that mass can be converted to energy, and vice versa, as given by

$$E = mc^2,$$

where E is energy, m is mass and c^2 is the square of the velocity of light.

A group of physicists has proposed a refinement of Einstein's relativity theory called doubly special relativity. The basic Einstein equation is revised to

$$E = \frac{mc^2}{1 + \dfrac{mc^2}{E_p}},$$

where E, m and c^2 are as described above, and E_p is what has come to be called Planck energy. D. Overbye, "E and mc^2: Equality It Seems, Is Relative," *The New York Times*, Dec. 31, 2002; F1.

What would have to be done to establish that doubly special relativity is a viable refinement of relativity? Explain.

15.8   References of Interest?

1. R. E. Peierls, *The Laws of Nature* (New York: Charles Scribner's Sons, 1956)

2. C. M. Will, *Was Einstein Right?* (New York: Basic Books, Inc., 1986)

16

Math Modeling for the Structure of Space

16.1   Preface

For more than two centuries after its formulation by Euclid (c. 330 B.C.)
Euclidean Geometry was viewed as the essence of space. Euclid's system,
however, contained a seed that led to the overthrow of this status and its
reduction to the more humble status of model of space.

These developments and their impact are the subject of the ensuing
discussion.

16.2   Euclidean Geometry

Geometry, and more generally mathematics, as a system of postulates,
definitions, and theorems or propositions deduced from the postulates, is the
great contribution of the ancient Greek mathematicians collectively. To any such
system they gave the name "Elements." Although not the first of its kind, the
most successful Elements were those compiled by Euclid of Alexandria around
300 B.C. [1]. Euclid set himself the task of taking the mathematics known at the
time and organizing it into a deductive system. This involved deciding on which
statements were to serve as the basic **assumptions** of the system (**postulates**),
which were to be theorems, and providing proofs for the theorems, original
ones when necessary. Euclid also had to decide how definitions were to be
formulated. It was an enormous undertaking and, considering the decisive
influence that his work has had, a brilliantly successful one.

Euclid took five geometric **postulates** as a basis for his deductive treatment of the geometry we now call Euclidean Geometry and introduced them in the following manner.

Let the following be **postulated**:

I.   To draw a straight line from any point to any point.

II.   To produce a finite straight line continuously in a straight line.

III.   To describe a circle with any center and distance.

IV.   That all right angles are equal to one another.

V.   That, if a straight line falling on two straight lines makes the interior angles on the same side less than two right angles, the two straight lines, if produced indefinitely, meet on that side on which the angles are less than two right angles. That is, in terms of Figure 16.1, if the sum of angles A and B is less than 180°, than lines N and M will intersect at some point on the same side as angles A and B.

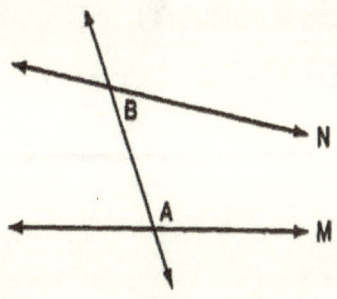

Figure 16.1

In more modern terminology, postulates I and II say: Through any two points, one straight line can be drawn; a line segment determines a line which is indefinite in extent. Postulate III says that a circle is determined when its center and radius are prescribed. Postulate IV says that all right angles are congruent. Postulates I-IV seem to express obvious truths in a simple fashion.

Postulate V, in contrast, is much more complex and as such stands in violation of the accepted criteria for postulates of the time, that they must express self-evident truths about spatial relations in a simple way. Euclid was aware of this, but part of his extraordinary accomplishment was to recognize that such a statement was needed to support the most complex of his geometric deductions and to boldly take it as a **postulate** after attempts to deduce it from simpler statements were unsuccessful.

Euclid's fifth postulate is known as his **parallel postulate**, although the term parallel does not occur in it. He defines parallel lines as lines which being in the same plane and being produced indefinitely in both directions do not meet. The label parallel postulate is appropriate because it is equivalent to the following statement which involves the term parallel, equivalent in the sense that it plus the other four Euclidean postulates imply Euclid's fifth postulate, and vice versa. This equivalent to Euclid's parallel postulate is the one usually cited as Euclid's parallel postulate in textbook expositions of Euclidean Geometry. It is known as Plairfair's postulate, after the Scottish physicist and mathematician John Plairfair (1748-1819) who popularized it in a very successful textbook that he wrote on *Euclid's Elements*.

If given a line L and point P not on L, then there is one and only one line which passes through P and is parallel to L (Figure 16.2).

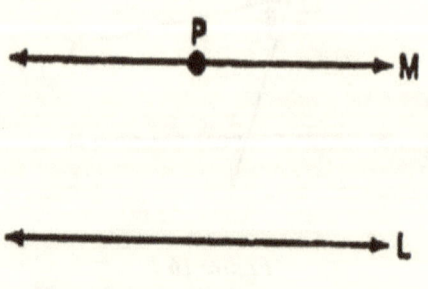

Figure 16.2

For the ancient Greek mathematicians and philosophers and their successors in mathematics, philosophy and science, Euclidean Geometry served as a description of the space in which we live and as an intellectual discipline.

It was not considered a mathematical model for space, with the admitted possibility that there might be other possible descriptions for space, but as the essence of space.

The afore sets in place the basic backround. In pursuing it to the development of non-Euclidean geometry and its impact on the math modeling perspective and mathematics more generally, it's useful to introduce what I term the parallel postulate problem.

16.3   The Parallel Postulate Problem

Euclid's contemporaries and successors greatly admired the organization of geometry which he had achieved in his *Elements*, but they were also dissatisfied with the price that had been paid in the form of a fifth postulate which could neither be considered simple nor self-evident.

The parallel postulate problem that arose was to free Euclid from this blemish by either deducing the fifth postulate from Euclid's other postulates, or replacing it with an equivalent postulate which was simple and self-evident. The problem attracted many scholars from many lands. Some attempts to solve it were ingenious, but all were unsuccessful.

In a departure from earlier approaches to the parallel postulate problem, which were direct, three mathematicians, working independently, brought an indirect approach to the problem. Each tried to show that Euclid's fifth postulate was a consequence of his other postulates by showing that if the fifth or its equivalent is replaced by its negation, the amended set of postulates has implications that are contradictory. None were successful. The results they obtained were in contradiction to the nature of space as it was then perceived, but they were not in contradiction with each other.

Lobachevsky

In the early nineteenth century the parallel postulate problem was taken up by three men who reached startlingly different conclusions from their predecessors. Nickolai Ivanovich Lobachevsky (1792-1856), of the then recently established Kazan University in southern Russia, was first to publicly announce and publish his results. Lobachevsky took the contradiction of Plairfair's equivalent of Euclid's fifth in the following form: If given a line

L and a point P not on L, there are at least two lines that pass through P and are parallel to L (see Figure 16.3). On the face of it this statement, which we shall term Lobachevsky's parallel postulate, seems absurd.

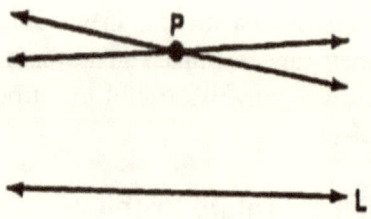

Figure 16.3

In deducing the consequences of the amended system consisting of Euclid's postulates, with his fifth replaced by his parallel postulate, Lobachevsky concluded that his system forms a consistent whole, free of contradictory statements, although it was strikingly at variance with the reality of space as it was then understood to be. He first outlined his ideas in a paper he presented at a meeting of the mathematics and physics division of Kazan University held on February 26, 1826. Three years later he elaborated on his ideas in his paper "On the Principles of Geometry," published in the Kazan Messenger. In the years 1835-1855 Lobachevsky further developed his non-Euclidean Geometry in a series of works [2].

Student Reaction

My students view Figure 16.3 as absurd. The idea of a system that is in contradiction to what they had been led to believe is the nature of physical space is ridiculous and not worth further attention. I remember being in the same boat many years ago (as were almost all of Lobachevsky's contemporaries), so that I can appreciate the difficulty they have in getting a grip on Lobachevsky's insight.

I know of no simple way for achieving understanding of Lobachevsky's insight. I can only recommend perseverance, which is what won the battle for me. Addressing the food-for-thought questions in sec. 16.8 is a helpful ally in achieving understanding, but they have to be given serious, thoughtful attention.

Like salmon who brave fierce currents to spawn, Lobachevsky was braving the fierce intellectual currents of his day by arguing that a system viewed as absurd, in contradiction to the reality of space, was internally consistent in its own right.

The hostility of the intellectual currents to Lobachevsky's ideas was substantially increased by the widespread acceptance in the late eighteenth and early nineteenth centuries of the views of the philosopher Immanuel Kant (1724-1804) on the nature of geometry.

Philosophical Blinders

Kant took the seemingly simple idea of simple, self-evident truth to the highest imaginable level of intellectual sophistication by endowing it with a thick layer of philosophical varnish.

He viewed the postulates and theorems of Euclidean Geometry as *synthetic a priori* propositions. Propositions whose truth or falsity can be shown by reason, prior to observation, are called *a priori* propositions. A proposition such as "All apples are fruit" is an *a priori* proposition, but it gives us no factual information; its truth follows from the meaning of the words. Propositions whose truth or falsity follows from the meaning of the words are called analytic propositions. "All apples are fruit" is an analytic *a* priori proposition. A proposition that has factual content, such as "Some politicians pass slippery figures to the public," is called a synthetic proposition.

Are there factual propositions whose truth can be established by pure reason, without recourse to observation? That is, are there synthetic *a priori* propositions? Kant argued [3] that the propositions of geometry and arithmetic are synthetic *a priori* propositions. They express factual statements, but they can be known by pure reason. The pro-Kantians believed that we have *a priori* knowledge of space; we do not know it from experience, but just the opposite; they are indispensable conditions to our having experience.

16.4   The Emergence of Non-Euclidean Geometry

Lobachevsky was appreciated by his colleagues as an outstanding teacher and administrator (having served as Rector of Kazan University from 1827-1846, and Assistant Guardian of the Kazan Educational District, (1846-1855), but his ideas on geometry were incomprehensible to them and were treated with tolerance at best and derision and ridicule at worst. One of Lobachevsky's papers came to the attention of Carl Friedrich Gauss (1777-1855), who appreciated its worth and had Lobachevsky elected a member of the Gottingen Scientific Society in 1842. In letters to friends Gauss expressed the highest praise for Lobachevsky's work, but he never gave it public support.

By the early 1820s Gauss had satisfied himself that a system based on a denial of Euclid's fifth together with Euclid's other postulates is consistent but, apart from sharing his ideas in letters with trusted friends, kept his views to himself. Gauss was a private person who shunned controversy and had no desire to subject himself and his ideas to the wrath of the pro-Kantians. As he put it in a letter to one of his friends he feared the "clamor of the Boeotians."

Lobachevsky's far reaching investigations into what we now call Lobachevskian Geometry convinced him that this system is mathematically legitimate in the sense of being consistent, that is, free of contradictory statements. While convincing, Lobachevsky's analysis fell short of being unequivocally conclusive on this point.

In 1868 Eugenio Beltrami (1835-1900) showed that Lobachevskian Geometry is as consistent as Euclidean Geometry by showing that if Lobachevskian Geometry contained an inconsistency, then an inconsistency must also exist in Euclidean Geometry.

A mathematical Samson seeking to topple Lobachevskian Geometry on the basis of inconsistency would also, if successful, topple Euclidean Geometry; Lobachevskian Geometry is thus as mathematically legitimate as Euclidean Geometry.

Showing that a less familiar, controversial structure is as structurally legitimate as a more familiar one brings the less familiar one closer to us. But it also raises another question. Is the more familiar structure, or perhaps we should say, seemingly more familiar structure, as consistent as the less familiar one? It has been shown that the answer is yes so that in the final analysis we can say that each is as mathematically legitimate as the other.

16.5   Euclidean Geometry Dethroned as the Essence of Space.

Since each is as consistent as the other, Euclidean and Lobachevskian Geometries can peacefully coexist as mathematical structures very nicely. But can they coexist as rival models of physical space? If, like the pro-Kantians, you believe in physical space as being synonymous with a unique set of principles, Euclidean principles, then the answer is no; from such a point of view there is only one way to do it, the Euclidean way, and that's that. The consistency of Lobachevskian Geometry in terms of its Euclidean counterpart makes it arbitrary, and thus indefensible, to chose one as the only possible model for space on the basis of intellectual comfort. An examination of the postulates or theorems of each geometry against the behavior of space in terms of experimentation and observation emerges as the only satisfactory means for settling the issue of which geometry, if either, is a realistic description of physical space.

This is the point of view opened up by the development of a non-Euclidean Geometry that is as mathematically respectable as Euclidean Geometry. In terms of our everyday experience—living and working in the limited space about us, Euclidean Geometry seems so "natural." But "natural" in terms of our immediate environment does not, and should not, be allowed to translate to "natural" and thus the only way to describe space beyond our limited experience.

In this connection, and more generally as well, our perspective is enhanced by an analogy due to Albert Einstein and his colleague Leopold Infeld:

> In our endeavor to understand reality we are somewhat like the man trying to understand the mechanism of a closed watch. He sees the face and the moving hands, even hears its ticking, but he has no way of opening the case. If he is ingenious he may form some picture of a mechanism which could be responsible for all the things he observes, but he may never be sure his picture is the only one which could explain his observations [4; p. 31].

The program that emerges entails assigning physical meanings to point and line and, in terms of these physical representations, empirically testing the realism of some key theorems of both geometries. An important property of Lobachevskian Geometry is that for "small regions" it differs little from Euclidean Geometry. For "sufficiently large" regions the differences become more substantial, but how large "sufficiently large" must be in order for significant differences to be revealed by experimental results is not clear.

One key result in both geometries concerns the interior angle sum of a triangle. In Euclidean Geometry this is 180 degrees. A possible approach suggested by this is to interpret point as the position of a celestial body, line segment as a beam of light projected from one such body to another, and determine the interior angle sum of large celestial triangles determined by three such celestial points. This is what Lobachevsky did in taking as his celestial points the Earth, Sun and star Sirius. He concluded that in this triangle the interior angle sum cannot differ from 180 degrees by more than 0.00000372 arc seconds. The result is inconclusive because we cannot answer the question of whether the distances involved in this celestial triangle are sufficiently large to reveal deviations from Euclidean Geometry.

Another complication, which Lobachevsky recognized, is that the geometry used to describe space goes hand in hand with properties of matter in space. It might be that a discrepancy observed in the interior angle sum of a triangle from 180 degrees could be explained by retaining the **assumptions of Euclidean Geometry** but at the same time modifying some **physical assumptions** involving mechanics or optics. By the same token the absence of a discrepancy might be compatible with the **assumptions** of non-Euclidean Geometry and suitable adjustments in our **assumptions** about the behavior of matter in space [5].

Space in the "Small"

As to the application of geometry to practical measurements, engineering, surveying, and the like, "in the small", as we say, Euclidean Geometry, suitably interpreted, is confirmed to a high degree of approximation as a theory of physical space. Lobachevskian Geometry "in the small" approximates Euclidean Geometry.

Many of its principles, though not all, may be considered physically confirmed to a high degree of approximation. Engineers, surveyors and scientists who use Euclidean Geometry in practice could use Lobachevskian Geometry instead, but it is simpler to use Euclidean Geometry since its formulas tend to be simpler.

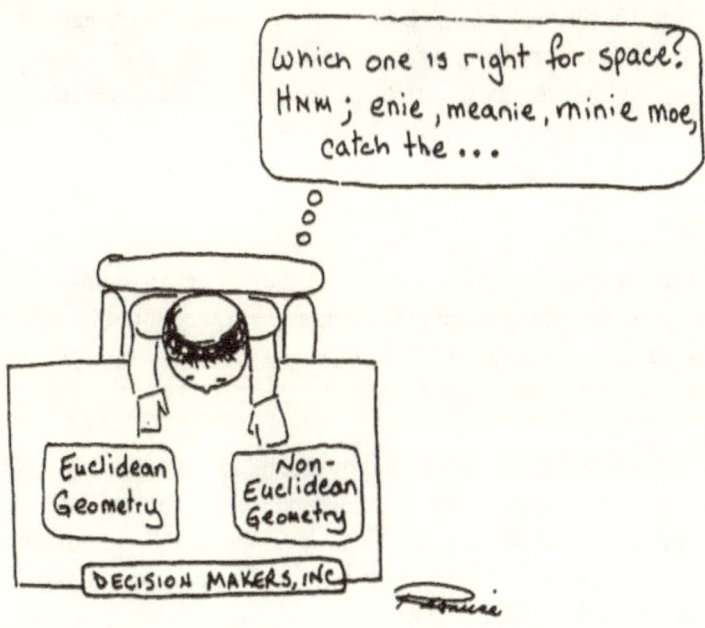

16.6   Impact: The "Big" Aftermath

> Yet it was coming; and through that criticism of first principles
> which Aristotle and Ptolemy and Galen underwent waited
> longer in Euclid's case than in theirs, it came for him at last.
> What Vasalius was to Galen, what Copernicus was to Ptolemy,
> that was Lobachevsky to Euclid . . . Each of them has brought
> about a revolution in scientific ideas so great that it can only be
> compared with that wrought by the other. And the reason of the
> transcendent importance of these two changes is that they are
> changes in the conception of the Cosmos [6].
>
>   W.K. Clifford

The English mathematician William Kingdon Clifford (1845-1879) expressed
this judgment in 1872 when the impact of non-Euclidean Geometry was
beginning to be felt. In an update Eric Temple Bell noted:

> The full impact of the Lobachevskian method of challenging
> axioms is probably yet to be felt. It is no exaggeration to call
> Lobachevsky the Copernicus of Geometry, for geometry is only
> part of the vaster domain which he renovated; it might even be
> just to designate him as a Copernicus of all thought [7].

Exaggeration or essence? This is not a simple question to come to grips with,
but to appreciate what prompted these appraisals we should look at the affect
that the creation of non-Euclidean Geometry has had on how subsequent
scholars view mathematics and science.

The geometric system perfected by Euclid came to occupy a position of
absolute authority for more than two thousand years. The rules Euclid laid down
for the geometry of space were seen as inviolate as the multiplication table.
Space and Euclid had become synonymous. The creation of non-Euclidean
Geometry forced a profound change in this point of view. The idea emerged
that other geometries, called non-Euclidean geometries, built from **postulates**

differing from those taken by Euclid, especially his parallel postulate, were not only logically possible, but might provide more realistic models of space.

> The fundamental ideas that an applied geometry is a model for space, not a "perfect" description of space, and that experimentation and observation provide us with the appropriate tools for deciding the issue of the realism of its propositions, are a direct consequence of the development of non-Euclidean Geometry. The general mathematical modeling point of view explored in this book is a development which emerged from this breakthrough.

The realization that **postulates** may be regarded as stated **assumptions** for purposes of further deduction, rather than supposedly self-evident truths, was slow in developing, but once it took hold it proved to be a tremendous liberating force. Mathematicians became free to study the implications of any **assumptions** that struck their fancy, no matter how fantastic and seemingly unrelated to the physical world they appeared to be, as long as they were internally consistent.

Another outcome of the successful launching of a non-Euclidean Geometry together with developments in the study of algebraic structures was that a new, more demanding level of deductive proof was brought to geometry in particular and mathematics in general, one which required that deductions stand on their own, independent of diagrams.

16.7   P.S. About Einsten's Model (ch. 15)

Space in the "Large"

Albert Einstein's Relativity Theory, based on a non-Euclidean Geometry, only became possible after the change in the intellectual climate initiated by the development of what we now call the classic non-Euclidean geometries (Lobachevsky—Gauss—Bolyai non-Euclidean Geometry being one).

16.8   Food for Thought

Carefully consider the following statements. If you Agree with a statement, explain the basis for your agreement; if you Disagree, explain the basis for your disagreement.

1. The parallel postulate problem was to show that Euclid's parallel postulate is true.

2. Lobachevsky sought to prove that Euclidean Geometry is inconsistent.

3. Immanuel Kant's philosophical views on the nature of mathematics provided a favorable climate for the development of non-Euclidean Geometry.

4. Since Lobachevskian Geometry contains statements that contradict statements in Euclidean Geometry, Lobachevskian Geometry cannot be considered a realistic description of physical space.

5. If Euclidean Geometry is not a perfect description of space, then the proofs of some Euclidean theorems must be in error.

6. Eugenio Beltrami showed that if Lobachevskian Geometry is consistent, then Euclidean Geometry must be inconsistent.

7. A mathematician proposed a geometry based on the postulates of Euclidean Geometry with two replacements.

Euclidean postulate	Replacement
1. Parallel postulate: Given a line L and point P not on it, there is one line passing through P parallel to L.	1. There are no parallel lines
2. Given two distinct points, there is one line containing them	2. Given two distinct points, there is at least one line (possibly more than one) containing them.

One argument brought against this proposed geometry is that it must contain contradictory statements (i.e., be inconsistent) because its parallel postulate (there are no parallel lines) and its second statement (two distinct points do not necessarily determine one and only one line) are contrary to the nature of physical space. Would you agree with this point of view?

8. One of the most important roles of proof in geometry is that it enables us to show that the postulates of geometry are true or false.

9. "The most suggestive and notable achievement of the last century is the discovery of Non-Euclidean Geometry." David Hilbert.

16.9   References of Interest?

1. For an English translation see T.L. Heath, *The Thirteen Books of Euclid's Elements*, 2nd ed., 3 vols (Cambridge Univ. Press, 1928; reprinted by Dover).

2. N.I. Lobachevsky, "Imaginary Geometry" (Scientific Papers of Kazan Univ., 1835); "Application of Imaginary Geometry to Certain Integrals" (Scientific Papers of Kazan Univ., 1836); "New Elements of Geometry with Complete Theory of Parallels" (Scientific Papers of Kazan Univ., 1835-38); *Geometrical Researches on the Theory of Parallels* (Berlin: Fincke, 1840). For an English translation see R. Bonola, *Non-Euclidean Geometry* (New York: Dover, 1955); *Pangeometry* (Scientific Papers of Kazan Univ., 1885); this work was dictated by Lobachevsky, then blind, in 1855, shortly before his death).

3. I. Kant, *Critique of Pure Reason*, 1781, 1787.

4. A. Einstein, L. Infeld, *The Evolution of Physics*, New Edition (New York: Simon & Schuster, 1961).

5. For further discussion see N. Daniels, "Lobachevsky: Some Anticipations of Later Views on the Relation between Geometry and Physics," *Isis*, vol. 66, no. 231 (March 1975), 75-85.

6. W.K. Clifford, "The Postulates of the Science of Space," Given before the British Association, 1872. Reprinted in J.R. Newman, *The World of Mathematics*, vol. 1 (New York: Simon & Schuster, 1956); 552-567.

7. E.T. Bell, *Men of Mathematics* (New York: Simon & Schuster, 1937); 306.

17

Which Approach to a Problem
is Most Suitable?

17.1 Preface

This question has, of course, come up in an earlier discussion of linear program modeling—The Austin Company's profit maximization problem (ch. 8) and Susan Reti's probability model problem (ch. 11), to take two examples—but another useful dimension to the question can be added by considering a fertilizer selection problem facing my hero of the story, Carl Cairns, a corn grower.

A new fertilizer has come to the market along with the usual claim that it's more effective than the one in current use for growing whatever it is that you grow. Carl wishes to test this claim by use of a statistical study and the basic question concerns the approach that would be most suitable. There are two options.

Option 1: Difference Between Means, Small Samples, Independently Chosen, approach

Option 2: Matched Pairs', Mean Difference, Small Samples, approach

Approach 1: Select 5 test plots at random from Carl's farm and treat them with the fertilizer in current use. Independently, choose another 5 test plots and treat them with the new fertilizer. Determine the yields (bushels per acre)

and examine the difference in sample means at some level of significance, 0.05, for example

Approach 2: Generate paired yields by selecting 5 plots at random, treat half of each plot with the fertilizer in current use and the other half of each plot with the new fertilizer. Determine the yields for each fertilizer and their mean values. Conduct an hypothesis test on the equalify of means arising from the data obtained in this way.

17.2   Which Hypothesis Test Approach is More Suitable?

Approach 1, the difference between means approach, does not effectively screen out extraneous factors (called **confounding factors**) which also play a role in determining crop yields. Such confounding factors include soil conditions, moisture, and weather conditions, which could vary considerably on the plots chosen (even though they are chosen at random) and mask the true effects of fertilizers on corn yields.

Approach 2, the matched pairs' mean difference approach, does more effectively screen out extraneous factors because the close proximity of the paired plots make it more unlikely that soil conditions, moisture, *et al* will vary significantly.

For Carl's situation the application of approach 1 to the null hypothesis of no difference between the population means of the fertilizer yields at the 0.05 significance level a test statistic which falls in the accept/reserve judgment region

What this result means to Carl is that there is no advantage to changing fertilizers in terms of increasing average corn yield.

The application of approach 2 yields a test statistic which falls in the reject region of the null hypothesis of no difference in the population means.

What this means to Carl is that the new fertilizer is more effective than the one in current use for growing corn.

In Carl's situation approach 2 is more effective in blocking out extraneous factors and more sharply focusing on the effects of the fertilizers. For discussion of details see [1; ch. 11, sec. 5 11.7-11.9].

Independently Drawn vs. Matched
Pair Samples: When Should they
Be Employed?

There is no sharp unequivocal answer to this question; it is a matter of judgment, but judgment refined through study and consideration of examples.

Situations involving a comparison between one way of doing things (old fertilizer, for example) and a new way (new fertilizer) where we want to block out extraneous effects and focus on the two ways might be handled advantageously through use of matched pair samples, depending on circumstances.

17.3   Confounding Factors

A fundamental question addressed in medical studies is, does factor A (diet, for example) cause (or prevent) factor B (heart disease, for example). Addressing this question is complicated by confounding factors which, if not taken into account, bias the conclusion obtained. This includes life style, state of health, exercise, environment, socio-economic status, *et al*. This situation provides an illuminating setting for obtaining insight into the complexities of dealing with bias due to confounding factors.

For discussion I recommend Gary Taubes' article [2].

17.4   Food for Thought Questions

Consider the following situations in which samples are to be drawn and a subsequent hypothesis test study concerning equality of means carried out. Should two samples be independently drawn (approach 1 for the subsequent study) or by the matched pairs (approach 2 for the subsequent study). Explain

1. To study the effectiveness of a new sleeping pill the sleeping patterns of a number of participants in a study are monitored before taking the pill and then after taking the pill. Their mean daily number of hours of sleep is determined before and after taking the pill.

2. A standard final examination in financial accounting is administered to all students in both the day and evening sessions at the Kaunas School of Business. To study whether there is a difference in the mean scores on the exam for the day and evening session students the chairperson of the accounting department plans to take samples of the day and evening session test scores and study the results.

3. A psychology project is being planned to study whatever differences exist between the IQ's of husbands and wives. Samples of men and their wives are to be chosen and their IQ levels determined and compared.

4. The Proctor Company makes light bulbs at their Waverly and Turin plants. The quality control manager wants to determine whether there is a difference in the mean lifetimes of the bulbs made by the two plants and plans to take samples from the output of both and analyze the results.

17.5   Reference of Interest?

1. W.J. Adams, Statistics: Basic Priciples and Applications, second edition, revised (Philaldelphia: Xlibris, 2009), available on the web at webpage. pace.edu/wadams.W.J. Adams, I. Kabus, and M. Preiss are the authors of the second edition.

2. G. Taubes, "Do We Really Know What Makes Us Healthy?" *The New York Times Magazine,* Sept. 16, 2007, pp. 52-59, 74, 78, 79.

18

Is Social Security on the Brink of Bankruptcy?

18.1   Preface

In what may be termed a five-year cycle for forecasting gloom and doom on Social Security George W. Bush's hand-picked commission on Social Security warned that 2016 is Social Security's crisis date—the year the program's obligations exceed its payroll tax revenues, so that it begins to run a deficit instead of a surplus. This time bomb must be defused, became the war cry, which opened the door to a number of proposals to save Social Security and the problem of saving Social Security from its would-be saviors.

What justifies the time-bomb image of Social Security? Mathematics does, it is argued, and sometimes "mathematics" is prefaced by simple or straightforward, and repeated in a tone which suggests that only a fool or simpleton would be willing to quarrel with mathematics, particularly when it's simple or straightforward.

18.2   What Are the Assumptions? Are They Realistic?

At the risk of being branded simpletons, we should be ready to quarrel with mathematics, not the technical part which comprises the deductive proof (unless it has a technical flaw), but in the sense of taking a close look at the **assumptions** that underlie the math deduction.

The fact is that the time-bomb projection made by The Advisory Council on Social Security (SSA) rests on what Aaron Bernstein of *Business Week* termed "**gloomy assumptions** about population and labor force growth that aren't shared by other government forecasters." "They're also difficult to square with historical trends," Bernstein noted [2], other key factors are the growth rates of gross domestic product (GDP) and productivity.

The average-annual growth rate of inflation-adjusted GDP and productivity over the next 75 years are **assumed** by SSA to be 1.4% and 1.3%, respectively, which are considerably gloomier than the 3.0% and 2.1% growth rates that we've had over the past 75 years. As Bernstein notes [3]. "if you buy the notion that we're in for 75 years of 1.4% growth, you have a lot more to worry about than just Social Security." Bernstein also points out that SSA has an alternate model in which GDP growth is **assumed** to average 2.14% a year over the next 75 years and that in this model Social Security is likely to show a slight surplus through 2072.

What becomes more and more clear, as a *Business Week* editorial pointed out: "A modest long-term problem is being hyped into an impending catastrophe. And battle lines drawn in Washington over Social Security have less to do with pragmatic solutions than with liberal-conservative wars dating back to the Roosevelt Administration [4].
       For discussion see [1]—[21].

18.3   Cooking Social Security's 'Deficit'

David Langer, *The Christian Science Monitor*, Jan. 1, 2000; 9

David Langer, a consulting actuary, is chairman of David Langer Company Inc., in New York.

I DISCOVERED in 1974 that cooking numbers is routine in Washington.

I had estimated that the new individual retirement account legislation would generate a federal tax loss of about $2.5 billion annually, while a congressional report cited a loss of only $200 million.

A senior IRS researcher laughed when I asked about it and said not to tell anyone. But that's how things are done: To promote legislation, the public will be given bargain-level cost estimates, while to scotch legislation, the reverse is true.

My curiosity was thus aroused several years ago by the widespread publicity coming from those wishing to privatize Social Security, who argued the program was going bankrupt. Saving it, they said, required that it be privatized and that benefits be substantially cut.

This concerned me deeply, since companies rely on Social Security as a base pension plan, as do most workers and their dependents. I therefore began an intensive investigation, reviewing the annual reports prepared by the trustees of the program's trust funds with the technical support of the Social Security Administration's actuaries.

I was surprised to learn that, contrary to the impression these actuaries give that they dictate the crucial economic and demographic **assumptions** underlying their financial projections, the choice is actually made by the *politically appointed* trustees.

Briefly, the actuaries provide the trustees with a preliminary report containing financial projections on the basis of recommended **assumptions** about the future (wage increases, gross domestic product, mortality rates, etc.), along with the projected deficit or surplus, for periods of up to 75 years.

The 75-year projections have yielded their largest deficits in recent years: more than 2 percent of the Social Security taxable wage base since 1994.

The trustees would then ask questions such as, "What if we lower the fertility rate by 5 percent?" And the senior actuary would then quickly estimate, based on rules of thumb, a deficit of, say, 2.2 percent. At some point, the trustees tell the actuaries the deficit level they desire. The actuaries will then put together the appropriate **assumptions and computations** for the trustees' annual report.

This procedure is, of course, improper, because it opens the system to political manipulation by the trustees. They are able, for instance, to make Social Security look as if it is in serious financial trouble and requires radical change to save it.

There is now sufficient evidence to conclude that this manipulation has, in fact, occurred.

A look at the actual projected 75-year deficits from 1980 to 1998 bears this out.

If one focuses only on those deficits, which stem mainly from changes in the **discretionary actuarial assumptions** and methods each year to meet the deficit goal of the trustees, one would find a continuous annual downward movement. The changes by 1998 added up to a total deficit of 2.29 percent. Compensating for this would require an increase in the payroll tax of about 1.15 percent each for workers and employers.

The explanations given by the actuaries in the annual reports have little credibility; they ignore considerable actual experience and rely, instead, on highly pessimistic speculation about the future. Also, of the 101 changes made in **assumptions** and methods from 1980 to 1998, 68 were negative and 33 positive, indicating a strong bias.

The Social Security actuaries also do not state in the trustees' annual report, as they are required to do, that key decisions are made by the trustees in what is basically an actuarial function.

The preliminary reports of the actuaries have been declared secret at a level reserved for the hydrogen bomb. We will thus never know how the trustee's deficit differs from the one initially calculated by the actuaries.

What motivates the powerful political push to privatize Social Security and cut its benefits? Briefly, the enormous potential wealth that privatization would bring to many financial institutions, who have contributed in excess of $50 million to campaigns. Further, cutting benefits raises the budget surplus—expected to reach a few trillion dollars—under the federal "unified budget." This reduces the national deficit and allows the government to spend that much more on favorite programs without the need to raise taxes.

The political trustees clearly had the motivation, opportunity, and means to advance the spurious concept of Social Security bankruptcy, and the evidence suggests they used their strategic position to further their goals.

From the public's point of view, Social Security has been thrust into the political arena in a way that is not desirable to workers and their dependents or to most employers. The solution is to avoid the accumulation of excessive reserves by reverting to a pay-as-you-go basis, under which surpluses are kept minimal and actuarial gamesmanship is more difficult.

The practice of determining what numbers you want and then generating the **assumptions** to obtain them is not unique to Social Security. The practice is called "backing in". See A. Berenson, "Tweaking Numbers to Meet Goals Comes Back to Haunt Executives", *The New York Times*, (June 29, 2002; A21) Berenson notes:

> Instead of first figuring out their sales and subtracting expenses to calculate the profit, they [some companies] work backward. They start with the profit that investors are expecting and manipulate their sales and expenses to make sure the numbers come out right.

18.4   References of Interest?

1. D. Baker, M. Weisbrot, *Social Security: The Phony Crisis* (Chicago: The University of Chicago Press, 1999; p.3).

2. A. Bernstein, "Social Security: Is the Sky Really Falling?, *Business Week*, Feb. 10, 1997.

3. A. Bernstein, "Social Security: Go Refigure," *Business Week*, Feb. 8, 1999.

4. Editorial: "What Social Security Crisis," *Business Week*, Nov. 30, 1998. This editorial was reproduced in *The New York Times*, under Business Week Commentary, Nov. 21, 1998 (A.15).

5. H. Aaron, "The Myths of the Social Security 'Crisis,'" *The Washington Post National Weekly Edition*, July 29, 1996.

6. P. Coy, "Social Security: Let It Be," *Business Week*, Nov. 30, 1998.

7. R. Eisner, *The Great Deficit Scares: The Federal Budget. Trade, and Social Security* (New York: The Century Foundation Press, 1997).

8. R. Eisner, *Social Security: More, Not Less,* (New York: The Century Foundation Press, 1998).

9. G. Koretz, "Stay Loose on Social Security," *Business Week*, March 29, 1999.

10. "Is There a Social Security Crisis? Sam Beard Debates Theodore R. Marmor & Jerry L. Mashaw," *The American Prospect,* Jan.-Feb., 1997.

11. J. Madrick, "Social Security and Its Discontents," *The New York Review of Books,* Dec. 19, 1996.

12. A. Munnell and R. K. Weaver, "Social Security's False Alarm", *The Christian Science Monitor*, July 19, 2001.

13. J. D. McKinnon, "Bush Commission Begins to Make Case That Social Security Must Be Overhauled", *The Wall Street Journal*, July 20, 2001.

14. R.W. Stevenson, "Panel Argues for Changing Social Security", *The New York Times*, July 20, 2001.

15. P. Krugman, "2016 and All That", *The New York Times*, July 22, 2001.

16. A. Goldstein, "Bitter Words over Social Security", *The Washington Post National Weekly Edition*, July 30-Aug. 12, 2001.

17. P. Krugman, "Nothing for Something", *The New York Times*, Aug. 8, 2001.

18. P. Krugman, "Fabricating a Crisis", *The New York Times*, Aug. 21, 2001.

19. D. Francis, "Social Security Short Falls are Suspect", *The Christian Science Monitor*, May 8, 2006.

20. D. Francis, "Social Security Faces a Crisis? Hardly", *The Christian Science Monitor*, Dec. 3, 2007; p. 17.

21. D. Francis, "Social Security: Sounder Than You Think", *The Christian Science Monitor*, April 7, 2008; p. 17.

19

Comments on Selected
Food for Thought Questions: 1

19.1   Chapter 1 (p. 44)

1. (a)   T1. Cost of cigarettes: $228,125

        T2. Insurance: 12,000

        T3. Health maintenance: 50,000

        T4. Total cost $290,125

    (b)   No; it's irrelevant. Warner and Wright's results are valid with respect to their respective **assumptions**.

    (c)   No; it's irrelevant to the validity issue.

    (d)   Not by itself; it depends on how realistic Warner and Wright were in taking into account the cost of cleaning of clothes, etc. in their models.

(e)   Yes, it serves as evidence that all three models are unrealistic in their estimates of the average cost of a pack of cigarettes over a fifty year period.

19.2   Chapter 2 (p. 53)

1. $\dfrac{350}{50} = 7$; No; it's irrelevant. The "proof" of Henry's theorem establishes its validity with respect to Henry's **postulates** and has nothing to do with questions involving Andy's actual trip time.

3.   Yes. To address the question of the realism of the valid conclusions obtained from Andy and Henry's models the trip would have to be taken along the routes for which these models were developed and a comparison between the deduced times of 6 hours (from Andy's model) and 7 hours (from Henry's model) made with the actual trip times.

   If the actual trip time for Andy's route is "close" to 6 hours, this would be evidence supporting the realism of Andy's model. If the actual trip time for Henry's route is close to 7 hours, this would be evidence supporting the realism of Henry's model.

6.   (a).   No. As noted in answer to the first question, the mathematical operation division, yielding $350 \div 50 = 7$, establishes the validity of Henry's conclusion from his **postulates**. Andy's trip time is irrelevant to the issue of validity.

   (b)   Andy's actual trip time of 5 hours and 55 minutes is "close" to the deduced trip time of 6 hours obtained from his model, which is evidence in support of the realism of his model.

8.   (a).   (i) Trip time: 5 hours, (ii) Trip time: 24 hours (iii) Trip time: 9 hours

   (b)   No; this information is irrelevant to the issue of validity.

   (c)   The actual trip time of 5 hours and 35 minutes is 10% off the deduced trip time of 5 hours, so that the difference between them is large enough that the times cannot be considered "close."

This establishes that Ann's model is not realistic for the first leg of the trip. The actual trip times for the second leg of the trip and the overall trip are "close" enough to the validly deduced trip times that they provide evidence supporting the realism of Ann's model for the second leg of the trip and the overall trip.

(d)   No. The validity of the theorems is irrelevant to the issue of their realism. The validity of the theorems is established by mathematical arguments. Their realism has to do with how accurate they are as descriptions of the real-world or institutional world that we have created.

19.3   Chapter 3 (p. 68)

1.   Valid; the hypothesis forces Joe Warren into the category of frogs.

3.   C1 is not valid; C1 is not a conclusion that is forced by the hypothesis.

C2 is not valid. C3 is not valid. C4 is not valid.

(a)   Mathematical conclusions are valid if they follow as logical consequences of the **assumptions,** irrespective of whether or not the **assumptions** are realistic.

(b)   If a valid conclusion is true, the assumptions may or may not be true.

Section 3.6 (p. 72)

9.   (a)   No; the validity of the predictions of quantum theory, outlandish or not, is a matter of whether they follow in a deductive-logical sense from the hypothesis of quantum theory. The creation of a Bose-Einstein condensate is irrelevant to this issue.

(b)   The creation of a Bose-Einstein condensate would help settle the question of realism of some of the "outlandish" predictions of quantum theory.

19.4 Chapter 4 (p. 85)

1. (a)   The corner-point method cannot guarantee that profit will be maximized when 300 ZKB-47 and 250 ZKB-82 units are made daily and sold. Whether these output levels or other ones will maximize profit is a question of realism, and the corner-point method, as a mathematical technique, can only ensure the **validity** of the predicted output levels with respect to the linear program model that was set up for the profit maximization problem in question.

     If the **assumptions** that the model reflects are **realistic**, then the afore output levels, obtained as a valid conclusion of these **assumptions**, will maximize profit; if these **assumptions are not realistic**, then it may well happen that other output levels would yield a higher profit than that projected for 300 ZKB-47 and 250 ZKB-82 units.

   (b)   The basis for implementing the conclusion that 300 ZKB-47 and 250 ZKB-82 units be made and sold daily is the judgment, based on an analysis of the company's operations and the market, that the **assumptions** made are realistic.

2. The realism of the team's postulates

19.5   Chapter 5 (pp 90)

1. The problem here is that what was actually proved is not the same as what was required to be proved. The proof given showed that an unique line *PE* parallel to *L* is obtained by the argument developed. This does not show that a line *M* parallel to *L* obtained from some other construction must the same as *PE*. Indeed, other constructions of parallel lines to *L* at *P* are known.

2. Proclus tacitly **assumes** that the distance between two parallel lines is bounded, which cannot be justified without the use of that which he is seeking to prove.

Section 5.5 (p. 95)

3. No; the argument does not establish that s and q are distinct.

4. No. Mr. Vlasik has six zogs in name, but not necessarily in number. He has to show that they are distinct.

5. No; that the argument given is 3 is invalid does not, by itself, exclude the possibility that there is a valid argument establishing 3.

7.

Proof:

Statement	Justification
1. Let p, q, and r denote blobs.	1. *P1*
2. Let R denote a *neighborhood* containing P but not q, and Q a *neighborhood* containing q but not p. `	2. *P3*
3. Let P denote a *neighborhood* containing p and q.	3. *P2*

4. P, Q, and R are distinct neighborhoods since they do not contain the same blobs.

19.6   Chapter 6 (p. 110)

3.

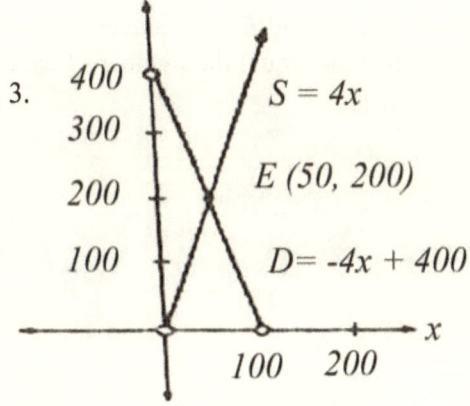

Figure 19.1

(c)   No; its math validity is established, but its real world accuracy
       hinges on the real-world accuracy of the demand and supply
       functions we are given.

19.7   Chapter 7 (p. 121)

1. Let x, y, and z denote the total costs of service departments S_1, S_2, and S_3,
    respectively, for March. The condition that the total cost of each service
    department must equal the overhead of the department (for March) plus
    the charges for services provided by the other service departments leads
    to the following relations:

$$x = 40{,}000 + 0.1y + 0.2z$$
$$y = 30{,}000 + 0.1x + 0.05z$$
$$z = 20{,}000 + 0.05x + 0.1y$$

By transposing terms we obtain the following system of equations:

$$x \quad - 0.1y \quad - 0.2z = 40{,}000$$
$$-0.1x \quad + y \quad - 0.05z = 30{,}000$$
$$-0.05x \; - 0.1y \quad + z = 20{,}000$$

2. We introduce variables s, t, u, v, w, and x as shown in the network diagram
    (Fig. 19.2). Variable x denotes the number of cars passing between points
    B and C per hour, v the number of cars passing between points C and E
    per hour, and so on.
        For equilibrium at the points A, B, C, D and E, the number of cars
    entering each of these points per hour must equal the number of cars
    leaving each of

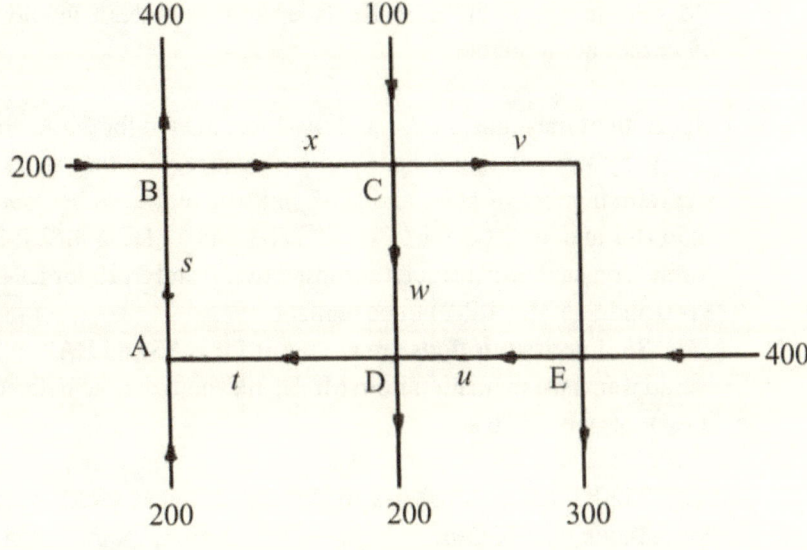

Figure 19.2

these points per hour. This leads to the following conditions:

A: $t + 200 = s$
B: $s + 200 = x + 400$
C: $x + 100 = v + w$
D: $u + \quad w = t + 200$
E: $v + 400 = u + 300$

Transposing and rewriting terms yields the following system of equations:

$$
\begin{array}{rl}
s - t & = 200 \\
s \quad\quad\quad - x & = 200 \\
v + w - x & = 100 \\
- t + u \quad\quad + w & = 200 \\
u \quad - v & = 100
\end{array}
$$

19.8   Chapter 8 (p. 136)

1. (a)   No; the latest computer technology, correctly applied, will give
us the valid conclusion that (30, 20) is the solution and 600 is the
maximum value and that's all. Its use does not address the issue of
the realism of the **assumptions**, which is our point of interest.

(b)   No; same answer as the afore. Replace latest computer technology by corner point method.

(c)   No; Both of the models LP-1 and LP-2 formulated for the Austin Company's production scheduling problem require that at least a certain number of DT-1 and DT-2 units be produced per week (25 DT-1 and 40 DT-2 for LP-1; 50 DT-1 and 50 DT-2 for LP-2) for the constant profit values **assumed** ($150 and $120 for LP-1; $140 and $150 for LP-2) to be realistic.

Similar assumptions are not made for RA5 and RA9 and I would want an explanation from Mr. Blank about why he believes it unnecessary to do so.

2. (a) and (b). Mathematics is precise in the sense that it yields valid conclusions with respect to an underlying starting point called a hypothesis. If there are two underlying hypotheses, then two quite different valid conclusions—solutions to linear program models in this case—might arise, each valid with respect to its own starting point.

(c)   What **assumptions** underlie M1 and M2? On what basis were they made? Convince me that they are realistic.

(d)   I would adopt that model based on what I judged to be realistic **assumptions**. If I were not convinced that the **assumptions** for either of M1 and M2 were realistic, I would not adopt either model. I would call for additional studies.

3. (a)   Different solutions were obtained on the basis of different **assumptions**. The mathematics is precise in the sense that its methods yields conclusions that are **valid with respect to the underlying assumptions**.

The fact that the simplex method was used in one situation whereas the corner-point method was used in the other instance is irrelevant as far as the validity of the conclusions obtained is concerned. Both procedures, as mathematical methods, yield valid conclusions.

(b)   Both solutions are correct in the sense of being **valid with respect to their underlying hypotheses**. Clearly both cannot be true since the maximum profit cannot be $3000 per week and at the same time $2500 per week.

(c)   The basic question is, which solution actually maximizes profit, or to allow a shade of gray, which solution comes closest to approximating the maximum profit? This is the solution that we wish to implement.

The **assumptions** from which these solutions follow as valid consequences should be carefully reviewed as to their realism. If the **assumptions** that lead to the solution (450,300), with a predicted maximum profit of $2500 per week, are more realistic than the **assumptions** that lead to the solution (500,280), with a predicted maximum profit of $3000 per week, then the (450,300) solution should be implemented since the predicted $2500 profit value will be closer to reality than the $3000 value.

Without question, the predicted $3000 profit is more appealing, but what is more appealing is not necessarily what is realistic. If the **assumptions** of both linear programs are **not realistic,** then neither one should be implemented.

19.9   Chapter 9 (p. 150)

1. We begin by introducing variables X_{11}, which relates candidate 1 (Ann) to job 1 (physiologist), X_{21}, which relates candidate 2 (Gena) to job 1 (physiologist), and so on. X_{11} can assume one of two values, 0 if candidate 1 is not assigned job 1 and 1 if candidate 1 is assigned job 1; X_{12} can assume one of two values, 0 if candidate 1 is not assigned job 2 and 1 if candidate 1 is assigned job 2; and so on. There are six variables (see Table 19.1) since there are 3 candidates and 2 jobs.

Table 19.1

Candidate		Job	
		Physiologist (job 1)	Biochemist (job 2)
	Ann (candidate 1)	X_{11}	X_{12}
	Gena (candidate 2)	X_{21}	X_{22}
	Marty (candidate 3)	X_{31}	X_{32}

The Vroman Institute's job-assignment problem leads to the following integer program: Find nonnegative integers (0's and 1's) which

$$\text{maximize } P = 3X_{11} + \frac{5}{2}X_{12} + 2X_{21} + \frac{5}{2}X_{22} + 2X_{31} + \frac{3}{2}X_{32}$$

subject to

$$
\left.
\begin{array}{l}
x_{11} + x_{12} \leq 1 \\
\quad\quad x_{21} + x_{22} \leq 1 \\
\quad\quad\quad\quad x_{31} + x_{32} \leq 1
\end{array}
\right\} \quad
\begin{array}{l}
\text{Each candidate is assigned to} \\
\text{at most one job.}
\end{array}
$$

$$
\left.
\begin{array}{l}
x_{11} \quad\; + x_{21} \quad\; + x_{31} \leq 1 \\
x_{12} \quad\quad + x_{22} \quad\quad\quad \leq 1 \\
\quad\quad\quad\quad\quad\quad\quad + x_{32} \leq 1
\end{array}
\right\} \quad
\begin{array}{l}
\text{Each job is filled by at most} \\
\text{one candidate.}
\end{array}
$$

2. To relate candidates 1 (Jones), 2 (Johnson), and 3 (Marks) to jobs 1 (Supreme Court Judge) and 2 (Civil Court Judge) we introduce variables X_{11}, X_{21}, and more generally X_{ij}, to relate candidate i to job j, as summarized in Table 19.2. X_{ij} can assume one of two values, 0 if candidate i is not assigned job j, 1 if candidate i is assigned job j.

Table 19.2

Candidate		Job	
		Supreme Court Judge (job 1)	Civil Court Judge (job 2)
Jones	(candidate 1)	X_{11}	X_{12}
Johnson	(candidate 2)	X_{21}	X_{22}
Marks	(candidate 3)	X_{31}	X_{32}

The problem of filling the positions so that the total potential rating is maximized in such a way that each candidate is assigned to at most one job and each job is filled by at most one person is expressed by the following 0-1 integer program:

$$\text{Max. } P = 9X_{11} + 8X_{21} + 10X_{31} + 8X_{12} + 9X_{22} + 8X_{32}$$

subject to

$$
\left.
\begin{aligned}
x_{11} + x_{12} &\leq 1 \\
x_{21} + x_{22} &\leq 1 \\
x_{31} + x_{32} &\leq 1
\end{aligned}
\right\}
\quad
\begin{aligned}
&\text{Each candidate is} \\
&\text{assigned to at} \\
&\text{most one job}
\end{aligned}
$$

$$
\left.
\begin{aligned}
X_{11} + X_{21} + X_{31} &\leq 1 \\
X_{12} + X_{22} &\leq 1 \\
+ X_{32} &\leq 1
\end{aligned}
\right\}
\quad
\begin{aligned}
&\text{Each job is filled} \\
&\text{by at most one} \\
&\text{candidate}
\end{aligned}
$$

3. Corresponding to sites $S_1, S_2, \ldots, S_5$, we introduce variables $X_1, X_2, \ldots, X_5$, respectively. X_1 can take on one of two values, 0 if site S_1 is not chosen and 1 if site S_1, is chosen; X_2 can take on one of two values, 0 if site S_2 is not chosen and 1 if site S_2, is chosen. Similarly for variables X_3, X_4, and X_5.

The function

$$P = 0.19X_1 + 0.23X_2 + 0.20X_3 + 0.30X_4 + 0.16X_5$$

obtained by multiplying the profit realized from a site by the variable associated with the site and adding, is the function to be maximized. When $X_1, \ldots, X_5$ are given values (zeros and ones), P becomes a sum of profits. For example, for $X_1 = X_2 = 1$ and $X_3 = X_4 = X_5 = 0$,

$$P = 0.19 + 0.23 = 0.42$$

That is, if sites S_1 and S_2 are chosen for development, the expected total profit is $0.42 million. The problem of selecting sites so that the total cost does not exceed the amount available ($32 million) and the total profit is maximized can be expressed in the following terms. Find nonnegative integers that:

Maximize $P = 0.19X_1 + 0.23X_2 + 0.20X_3 + 0.30X_4 + 0.16X_5$
subject to

$$X_1 \leq 1$$
$$X_2 \leq 1$$
$$X_3 \leq 1$$
$$X_4 \leq 1$$
$$X_5 \leq 1$$
$$10X_1 + 12X_2 + 11X_3 + 15X_4 + 9X_5 \leq 32$$

The nonnegative statement and the first five inequalities express the requirement that the integer values of the variables be zeros or ones. The last inequality expresses the requirement that the total amount invested not exceed $32 million.

19.10   Chapter 10 (p. 186)

2 (a) $I = \dfrac{5.51}{4} \cdot 100 = 138$

(b) $I = \dfrac{137}{103} \cdot 100 = 133$

3. (a) (i) $I = \dfrac{2.72}{1.53} \times 100 = 178.$

   (ii) $I = \dfrac{4.624}{4} \times 100 = 116.$

(b) $I = \dfrac{9,482}{4,993} \times 100 = 190.$

(c) $I = \dfrac{19,040}{9,482} \times 100 = 201.$

(d) $I = \dfrac{8,494}{4,646} \times 100 = 183.$

(e) Same answer as (d).

4. (a) 106

(b) $\dfrac{\$85}{1.06} = \$80.19$

(c) Yes; $0.19; more buying power.

5. (a) $I = \dfrac{8.47}{113.4} \cdot 100 = \$7.47, I = \dfrac{6.12}{113.4} \cdot 100 = \$5.40, I = \dfrac{9.78}{113.4} \cdot 100 = \$8.62$

   for service, retail trade and manufacturing, respectively. Relative to 2000, a typical 2004 worker in the retail trade industry has less purchasing ability, while a typical 2004 worker in each of the other industries has more purchasing ability in 2004.

(b) $I = \dfrac{8.81}{113.8}(107.3) = \$7.88,\ I = \dfrac{6.27}{120.0}(107.3) = \$5.61,$ and

   $I = \dfrac{10.77}{120.0}(107.3) = \$9.63,$ respectively. Relative to 2002, a typical

   2006 worker in the service industry has virtually the same

purchasing power, while a typical 2006 worker in each of the other industries has less purchasing ability.

(c) $I = \dfrac{8.81}{120.0}(113.4) = \$8.33$, $I = \dfrac{6.27}{120.0}(113.4) = \$5.93$ and

$I = \dfrac{10.77}{120.0}(113.4) = \$10.18$, respectively. Relative to 2004, only average wages in the manufacturing industry rose at a faster rate than did average prices during the 2004-2006

6. (a) $I = \dfrac{5916.7}{109.8} \cdot 100 = 5388.6$, $I = \dfrac{6553.0}{116.5} \cdot 100 = 5624.9$, and

$I = \dfrac{7253.8}{122.9} \cdot 100 = 5902.2$.

(b) (i) $I = \dfrac{5624.9 - 5438.7}{5438.7} \cdot 100 = 3.4\%$

(ii) $I = \dfrac{5902.2 - 5388.6}{5388.6} \cdot 100 = 9.5\%$

(iii) $I = \dfrac{5902.2 - 5438.7}{5438.7} \cdot 100 = 8.5\%$

(c) Percentage growth rates of actual GDP are:
$\dfrac{6553.0 - 5438.7}{5438.7} \cdot 100 = 20.5\%$, $\dfrac{7253.8 - 5916.7}{5916.7} \cdot 100 = 22.6\%$,

$\dfrac{7253.8 - 5438.7}{5438.7} \cdot 100 = 33.4\%$. The portion of the growth rate of

actual GDP that expresses real growth in GDP is: $\dfrac{3.4}{20.5} \cdot 100 = 16.6\%$,

$\dfrac{9.5}{22.6} \cdot 100 = 42\%$, and $\dfrac{8.5}{33.4} \cdot 100 = 25.4\%$. Therefore, 83.4%,

58.0% and 74.6%, respectively, of actual growth in GDP may be attributable to inflation.

7. (a) $\dfrac{\$2.10}{107} \cdot 100 = \$1.96$. No, since $\$1.96 \neq \$1.80$.

    (b) $\dfrac{\$2.00}{112} \cdot 113 = \$2.02$. No, since $\$2.02 \neq \$2.30$.

    (c) $\dfrac{\$2.00}{112} \cdot 100 = \$1.79$. No, since $\$1.79 < \$1.80$.

8. (a)   (i) $No;\ \dfrac{3}{80} = 0.0375 = 3.75\% < 5.6\%$.

            (ii) $Yes;\ \dfrac{15}{80} = 0.1875 = 18.75\% > 15.3\%$.

    (b) $For\ 2002, \dfrac{\$83}{1.056} = \$78.60.$ For 2004, $\dfrac{\$95}{1.153} = \$82.39.$

    (c)   (i) $No;\ \dfrac{1}{10} = 0.1 = 10\% > 5.6\%$.

            (ii) $No;\ \dfrac{3}{10} = 0.3 = 30\% > 26.6\%$.

    (d) $\dfrac{126.6 - 115.3}{115.3} = 0.098 = 9.8\%,\ \dfrac{\$53}{1.098} = \$48.27;$ since

    (e) $\dfrac{53 - 51}{51} = 0.039 = 3.9\% < 9.8\%$, the fee increase was not in line.

            $\dfrac{126 - 105.6}{105.6} = 0.199 = 19.9\%,\ \dfrac{\$110}{1.199} = \$91.75.$

            $\dfrac{126 - 115.3}{115.3} = 0.098 = 9.8\%.\ \dfrac{110 - 95}{95} = 0.1579 =$

            15.8%. Since $9.8\% \neq 15.8\%$, the legal fee increase was not in line.

19.11   Chapter 11 (p. 207)

1. (a) $P(E) = P(2) + P(4) + P(6) = 3/4 = 0.75$. (b) If a die whose behavior
         is described by model Y3 is tossed a large number of times, an even
         number will show approximately 75% of the time. (c) The tossing
         of the die a large number of times and observing how often an even
         number shows is **irrelevant** to the **validity issue. The validity** of
         $P(E) = 0.75$ was established by adding up the probabilities with
         which 2, 4 and 6 show in model Y3 and obtaining 0.75.

   (c)   From tossing the die a large number of times, we have that the
         relative frequency with which an even number showed is 0.665,
         which is markedly at variance with the predicted relative frequency
         of approximately 0.75 in (b). This establishes that $P(E) = 0.75$,
         interpreted in relative frequency terms, is **false**. (e) Model Y3 is
         not realistic for the die in question since **a valid conclusion of
         the model has been shown to be false.**

3. (p. 211)

   (a)   Jack's **assumption**, that the sample points in his sample space are
         equally likely to occur, is **not an accurate reflection** of the fairness
         of the dice. Consider, for example, the sample points {1,1} and
         {1,2}. {1,1} can only occur in one way, both dice must fall with
         1 showing. {1,2} can occur in two ways, the first die shows 1 and
         the second die shows 2, the first die shows 2 and the second die
         shows 1. Thus for fair dice it is reasonable to expect that over the
         long run the event {1,2} will occur approximately twice as often
         as the event {1,1}.

         Thus it would be more realistic to assign {1,2} a probability which
         is twice that assigned to {1,1}, instead of assigning both the same
         probability as Jack has done.

   (b)   Let x denote the probability to be assigned to {1,1}. The sample
         points {2,2}, {3,3}, {4,4}, {5,5}, {6,6} should all be assigned
         the **same probability x since they all can only occur in one
         way.** The remaining sample points, {1,2}, {1,3}, {2,3}, and so

on, should all be assigned a probability **2x since each can occur in two ways.** We have six sample points that will be assigned a probability value x, yielding a sum of $6x$, and fifteen sample points that will be assigned a probability value $2x$, yielding a sum of $30x$. The total sum of the probabilities of the sample points is thus $36x$. Since this total sum must be equal to 1 we have

$36x = 1$, which yields $x = \dfrac{1}{36}$.

Thus we are led to define P on Jack's sample space by $P(\{1,1\}) = P(\{2,2\}) = P(\{3,3\}) = P(\{4,4\}) = P(\{5,5\}) = P(\{6,6\})$ = 1/36, and the probability 2/36 for each of the remaining sample points—$\{1,2\}$, $\{1,3\}$, $\{2,3\}$, and so on.

5. (a) (1) Mark's probability function is **not realistic** because it does not take into account the different proportions of good bulbs made by plant $P1$, good bulbs made by plant $P2$, etc. From the

data given the proportion of good bulbs (of the total output of 8000 bulbs) made by $P1$ is $\dfrac{4950}{8000}$, the proportion of defective bulbs made by $P1$ is $\dfrac{50}{8000}$, etc.

(2) Mark's conclusion is **valid with respect to his model,** but **not realistic** in terms of the process of choosing a bulb at random from the day's output with the given proportions of good and defective bulbs.

(b) Bob's model is open to the same sort of criticism leveled at his brother's model.

(c) Yes, as follows: $P(GP1) = \dfrac{4950}{8000}$, $P(GP2) = \dfrac{2985}{8000}$, $P(DP1) = \dfrac{50}{8000}$, $P(DP2) = \dfrac{15}{8000}$.

(d)   Yes: $P(G) = \dfrac{7935}{8000}$, $P(D) = \dfrac{65}{8000}$.

6. (a)   No. This relative-frequency information has **no bearing on the validity of Andrew's conclusion**. The validity of Andrew's conclusion follows from the probability model that he set up.

(b)   No. The discrepancy between Andrew's valid conclusion interpreted in relative frequency terms (a head and tail will show approximately 33.3% of the time) and what actually happened (a head and tail showed 49.2% of the time) shows that Andrew's conclusion is **false about the behavior of well-balanced coins.**

(c)   Andrew's probability model is not a realistic model for describing the behavior of well-balanced coins.

(d) $S = \{HH, HT, TH, TT\}$, $P\,(HH) = \ldots = P(TT) = \dfrac{1}{4}$.

7. (a)   Consider the sample point (i,j). If $i+j$ is even (such as $(1,1)$, $(1,3)$, etc.) $P(i, j) = \dfrac{2}{54}$; if $i+j$ is odd (such as $(1,2)$, $(1,4)$, etc.), $P(i, j) = \dfrac{1}{54}$.

(b)   (i) $P(even\,sum) = \dfrac{36}{54}$; (ii) $P(i + j = 7) = \dfrac{6}{54}$; (iii) $P(i+j<6) = \dfrac{14}{54}$.

8. (a)   Jack's conclusion is **valid with respect to his probability model;** it follows as an inescapable consequence of his model. To establish this we note that

$$P \text{ (head shows twice in the three tosses)} = P(2) = \frac{1}{4}$$

(b)   Jack's conclusion, interpreted in relative-frequency terms, is that if a well-balanced coin is tossed three times in succession a large number of times, head will show twice in the three tosses approximately 25% of the time. **This conclusion was proved false** by the data collected in repeating the process a large number of times; head was observed to show twice in the three tosses 38% of the time, which is a considerable deviation from the predicted 25% value.

(c)   Jack's conclusion, **while valid** with respect to his probability model, **is false** about the behavior of a well-balanced coin. The discrepancy between the predicted 25% value and the actually observed 38% value can be traced back to unrealistic **assumptions** about the coin that Jack's probability model reflects. Jack is **assuming** that all of the sample points in his probability model have the same likelihood of occurrence. But it is unrealistic to **assume**, for example, that 0 heads in three tosses is as likely to occur as 1 head in three tosses when a well-balanced coin is tossed three times in succession. 0 heads can only be realized in one way, when you have TTT—tail followed by tail followed by tail; 1 head can be realized in three ways—HTT, THT and TTH. With a well-balanced coin one would expect 1 head to show approximately three times as often as 0 heads, and this feature is not realistically reflected by the same assignment of probabilities to these sample points.

9. (a)   From the nature of the question itself it is clear that Herman sees mathematics as a precise subject in the sense that its conclusions are truths about the dice tossing process under consideration. He is surprised because he has painfully encountered a conclusion obtained by mathematical reasoning (and interpreted in relative-frequency terms) that is false. Herman's faith in the *precision* of mathematical reasoning is misplaced. Mathematical reasoning is precise in the sense that its conclusions are valid with respect to the **assumptions** made; valid conclusions are not necessarily true statements about the process in question.

(b)   Herman's conclusion is **correct only in the sense of being valid with respect to his probability model and the assumptions that it**

reflects; if we accept his probability model as a point of departure, then we must accept his conclusion as following from it.

(c)   What went wrong is that the **probability model** that came with Herman's dice **was not a realistic reflection of the nature of his dice**. Herman's probability model is a good fit to well-balanced dice, and he had a pair of "loaded" dice.

12. (a)   Al's conclusion is **valid** in that it follows as an inescapable consequence of his probability model.

$$P(B) = P(G_1 G_2) = \frac{1}{4}$$

(b)   If the sampling procedure is repeated a large number of times—that is, two items are drawn at random from a lot of twenty where the first item drawn is not replaced before the second is drawn, and this is done a large number of times—then the sample drawn will consist of two good items approximately 25% of the time.

(c)   In performing the underlying process 300 times the following was obtained:

$$Relative\ freq.\ of\ B = \frac{238}{300} = 0.79$$

Event B was found to occur 79% of the time, which differs markedly from the 25% value obtained from Al's probability model.

(d)   As established in part (a), Al's conclusion that the probability that the sample drawn consists of two good items is 0.25 follows as an inescapable consequence of his probability model, and thus is valid with respect to his model by virtue of the meaning of validity. **The data cited in part (c) has no bearing on the validity of his conclusion.**

(e)   The data cited in part (c) establishes that Al's conclusion, while valid with respect to his model, is false about the underlying process when interpreted in relative-frequency terms. The appearance of a valid conclusion that differs so strikingly from the results obtained by performing the process a large number of times establishes that Al's model is not a realistic one for the inspection procedure.

(f)   For convenience of discussion let us think of the items in the lot as labeled I_1, I_2, . . . , I_{20}. Another approach to the study of the sampling process is to take as our sample space S the events expressed by all combinations of 2 items that can be drawn from the lot of 20 items.

$$S = \{(I_1,I_2), \quad (I_1,I_3), \ldots, (I_{19},I_{20})\}$$

The probability function P that best reflects the randomness of the selection procedure is the one that assigns the same value,

$$\frac{1}{C(20,2)} = \frac{1}{90},$$ to each sample point in S.

$$P(I_1,I_2) = \ldots = P(I_{19},I_{20}) = \frac{1}{90}$$

(g)   As a **valid conclusion** of the probability model developed in (f) we obtain the following.

$$P(2 \text{ good items}) = \frac{C(18,2)}{C(20,2)} = \frac{153}{190} = 0.805$$

The relative-frequency interpretation of this conclusion is that if the sampling procedure is repeated a large number of times, the sample drawn will consist of two good items in the neighborhood of 80.5% of the time.

(h)   The 79.3% value obtained by repeating the sampling procedure a large number of times is in close agreement with the 80.5% value obtained as a valid consequence of the probability model developed in (f). This provides **support for the realism of this model** for the sampling procedure.

13. (a)   Sample space S is the collection of events expressed by all combinations of 500 fish that can be selected from N fish. There are C(N,500) sample points in S. **Basic assumption;** The 300 fish that were initially caught, tagged, and release into the lake, were well-dispersed in the lake before the second sample of 500 fish were caught. We define P to be the function that assigns the same value, $\dfrac{1}{C(N_1 500)}$, to each sample point in S.

(b)   The number that maximizes the probability of catching 2 tagged fish in the batch of 500 that were caught.

(c) $\dfrac{C(300,\ 20)\cdot C(N-300,\ 498),}{C(N,500)}$ (d) 75,000

(e)   Yes; if our basic **assumption** is unrealistic, the realism of the 75,000 estimate is open to question and it may be much too low or much too high.

Section 11.8 (p. 250)

16. **Bernoulli trial model** defined by: $n = 1,000$, $E =$ event that any one of the light bulbs is defective, $p = 0.01$, $q = 0.99$. **Assumption:** Whether or not any bulb is defective is independent of whether or not any of the other bulbs are. A success is the occurrence of a defective bulb. $X =$ the number of defective bulbs from the 1,000.

$P(X \leq 15) = P(X = 0) + P(X = 1) + \cdots + P(X = 15) =$
$C(1000, 0)(0.01)^0(0.99)^{1,000} + C(1000, 1)(0.01)(0.99)^{999} + \cdots$
$+C(1000,15)(0.01)^{15}(0.99)^{985}$.

18. **Bernoulli trial model** defined by: $n = 10,000$, $E =$ event that an allergic reaction develops, $p = 0.005$, $q = 0.995$. **Assumption:** Whether or not one person develops an allergic reaction is independent of whether or not another person develops one. $X =$ the number of people out of the 10,000 that develop an allergic reaction.

$P(X < 100) = P(X = 0) + P(X = 1) + \cdots + P(X = 99) = C(10000, 0)(0.005)^0(0.995)^{10,000} + C(10000,1)(0.005)(0.995)^{9,999} + \cdots + C(10000,99)(0.005)^{99}(0.995)^{9,901}$.

19. **Bernoulli trial model** defined by $n = 10,000$, E is the event that a teenager has the measles, $p = 0.002$, $q = 0.998$, **Assumption**: Whether or not one teenager contracts measles or not is independent of whether any other teenager contracts measles or not.

   (a)   Let X = number of successes (number of teenagers who contract measles). $P(X \leq 25) = 0.8907$ (normal curve estimate)

   (b)   If a large number of groups of 10,000 teens are selected, in approximately 89% of the cases at most 25 teens will have the measles.

   (c)   No; since measles is a highly contagious disease, the independence condition is seriously compromised as is the realism of the numbers that follow as valid consequences of this **assumption**

20. B.   The blood test indicates the absence of the disease (test -). A_1: A person with the disease is selected (person +). A_2: A person not having the disease is selected (person -). From the tree graph of Figure 19.3 we have:

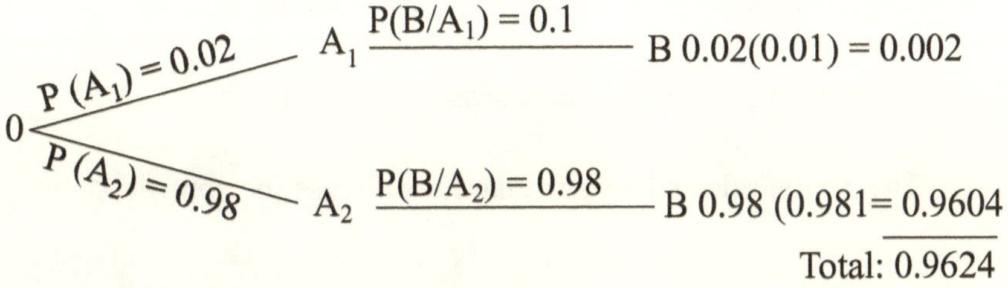

$$0 \quad \begin{array}{l} P(A_1) = 0.02 \quad A_1 \quad \dfrac{P(B/A_1) = 0.1}{} \quad B\ 0.02(0.01) = 0.002 \\[2em] P(A_2) = 0.98 \quad A_2 \quad \dfrac{P(B/A_2) = 0.98}{} \quad B\ 0.98\ (0.981 = 0.9604 \end{array}$$

Total: 0.9624

Figure 19.3

   (a) $$P(A_1 / B) = \frac{0.0020}{0.9624} = 0.002.$$

(b)   In the long run, approximately 0.2% of the people testing negative will turn out to have the disease.

(c)   1.   In the population under consideration 2% have the disease in question.

2.   Of those who are + (in terms of the standard test being used), 90% of the blood test examinations were +.

3.   In terms of the standard test being used, of those who were -, 98% of those given the blood test were correctly diagnosed as -.

4.   **Postulate P.** Other factors that may bear on the reliability of the blood test are considered negligible.

(d)   In the sense of being a valid consequence of the afore **assumptions**, yes; in the sense of being real world realistic, this issue is beyond its scope.

(e)   Yes; if the afore **assumptions** are unrealistic, this may be the case.

21. B_1: The person selected tests positive for the disease (test +).

B_2: The person selected tests negative for the disease (test -).

A_1: The person selected has the disease (person +).

A_2: The person selected does not have the disease (person -)

(a)   From Bayes's theorem we have: $P(A_1 / B_1) = \dfrac{0.0294}{0.0391} = 0.75$

(b)   From the tree graph of Figure 19.4 we have:

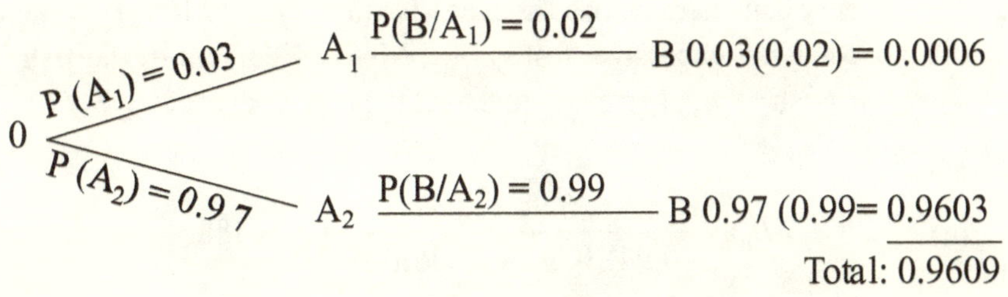

$$P(A_1) = 0.03 \quad A_1 \quad \frac{P(B/A_1) = 0.02}{} \quad B\ 0.03(0.02) = 0.0006$$

$$0$$

$$P(A_2) = 0.97 \quad A_2 \quad \frac{P(B/A_2) = 0.99}{} \quad B\ 0.97\ (0.99 = 0.9603$$

Total: 0.9609

Figure 19.4

$$P(A_2 / B_2) = \frac{0.0006}{0.9609} = 0.0006.$$

(c)   In the long run, approximately 75% of the people chosen at random testing + for the disease will actually have it and 0.06% of the people testing - for the disease will actually have it.

(d)   1.   In the highly populated industrial center under consideration 3% have lung cancer.

2.   Of those who are + (in terms of the standard test being used), 98% of those tested by the X-ray detection procedure were +.

3.   In terms of the standard test being used, of those who were -, 99% of those tested by the X-ray detection procedure tested -.

4.   **Postulate P:** Other factors that may bear on the reliability of the X-ray procedure are negligible.

(e)   In the sense of being a valid consequence of the afore **assumptions**, yes; in the sense of being real world realistic, this issue is beyond its scope.

(f)   Yes; if the afore **assumptions** are unrealistic, this may be the case.

22. B_1 : Mercury content diagnosed as excessive (test +). B_2 : Mercury content diagnosed as not excessive (test -). A_1 : Mercury content of the fish is excessive (fish +). A_2 : Mercury content of the fish is not excessive (fish -).

(a) $$P(A_2 / B_1) = \frac{(0.94)(0.04)}{(0.94)(0.04)+(0.06)(0.99)} = 0.388 .$$

(b) $$P(A_1 / B_2) = \frac{(0.06)(0.01)}{(0.06)(0.01)+(0.94)(0.96)} = 0.00066 .$$

(c)   In the long run, approximately 38.8% of the fish diagnosed as having excessive mercury content (test +) will actually not have it (fish -) and 0.066% of the fish diagnosed as not having excessive mercury content (test -) will have it (fish +).

(d)   1.   In Lake Bennett 6% of the fish population are + (have excessive mercury content).

       2.   In terms of the standard test being used, of those who are +, 99% of those tested by means of the new test were +.

       3.   In terms of the standard test being used, of those who were -, 96% of those tested by the new test were -.

       4.   Postulate P. Other factors that may bear on the reliability of the new test are negligible.

(e)   In the sense of being a valid consequence of the afore **assumptions**, yes; in the sense of being real world realistic, this issue is beyond its scope.

(f)   Yes; if the afore **assumptions** are unrealistic, this may be the case.

23. B: 7,000 or more units were sold. A_1 : The product is a big seller. A_2 : The product is a fair seller. A_3 : The product is a poor seller.

$$P(A_3 / B = \frac{0.04}{0.61} = 0.07.$$

24. B: More than 200 units were sold, A_1 : The computer is a big seller, A_2 : The computer is a good seller, A_3 : The product is a poor seller.

(a) $P(A_1 / B) = 0.726$;   (b) $P(A_2 / B) = 0.242$;

(c) $P(A_3 / B) = 0.032$;   (d) $0.726 + 0.242 = 0.968$.

(e)   No; Bayes's theorem is irrelevant to the issue of realism. Beware the **assumptions**; review the subjective probabilities that were assigned.

25. **Relative-frequency interpretation**: if the underlying process is repeated a large number of times, event E will occur in the neighborhood of 80% of the time.

   Subjective probability interpretation: 0.80 is a numerical expression of some individual's degree of belief in the occurrence of event E in connection with the underlying process.

   The relative-frequency interpretation presupposes that the process can be repeated a large number of times and has an objective content that is independent of the observer. The subjective probability interpretation can be entertained in connection with once-and-only situations. The subjective probability assigned to an event depends very much on the observer.

27. Both probabilistic statements can only be interpreted in **subjective** terms. Both are numerical expressions of degree of belief connected with once-and-only situations.

29. **Subjective probability terms**; 0.95 is a numerical expression of Eric's degree of belief that he will get an A in Sociology this semester.

19.12   Chapter 12 (p. 275)

1. row 1, column 1, value 1

3. row 1, column 1, value 2

4. maximin = minimax = 1

5. (a)

Colomn

Corporation

C_1 C_2 (off –

(*center*) (*center*)

			C_1 (center)	C_2 (off-center)
Row	(center)	R_1	0.5	0.65
Corp.	(off center)	R_2	0.4	0.5

Optimal strategies: row 1, column 2, value 0.5.

(b)   Not necessarily. This is a question of realism, which hinges on the realism of the **assumptions** made.

6. Since maximin = -1 while minimax = 1, the game must be analyzed in probabilistic terms. The row player's expected values with respect to columns 1 and 2 being played by the column player are

$$y_1 = 2p - 1, \quad y_2 = -2p + 1$$

respectively, where p is the probability that the row player plays now 1. Setting y_1 equal to y_2 and solving for p yields p = 1/2. Thus the row player's optimal strategy is to play both rows 1 and 2 with probability 1/2. The value of the game is 0, the common value of y_1 and y_2 for p = 1/2.

The column player's expected values with respect to rows 1 and 2 being played by the row player are

$$z_1 = -2r + 1, \quad z_2 = 2 - 1$$

respectively, where r is the probability that the column player plays column 1. Setting z_1 equal to z_2 and solving for r yields r = 1/2. Thus the column player's optimal strategy is to play both columns 1 and 2 with probability 1/2.

7. Play rows 1 and 2 with probabilities $\frac{1}{5}$ and $\frac{4}{5}$, respectively; play columns 1 and 2 with probabilities $\frac{2}{5}$ and $\frac{3}{5}$, respectively; value $\frac{18}{5}$.

9. (a)

		Market	
		Not over 40	Over 40
		C_1	C_2
Beverage (taste)	R_1	0.5	0.3
Beverage (low calorie level)	R_2	0.25	0.6

Optimal strategy; play rows R_1 and R_2 with probabilities $\frac{7}{11}$ and $\frac{4}{11}$, respectively; that is, stress taste in $\frac{7}{11}$ or 64% of the advertisements and stress low calorie level in $\frac{4}{11}$ or 36% of the advertisements. The value of the game is $\frac{9}{22}$ or 0.41; this is interpreted to mean that, with respect to the stated **assumptions**, the advertising mix will be effective on approximately 41% of the viewing audience.

(b)   Not necessarily. As with Food for Thought Question 5, this is a question of realism, which hinges on the realism of the assumptions made.

19.13   Chapter 13 (p. 284)

1. (a) $A = \begin{bmatrix} 0.2 & -0.3 \\ 0.3 & 0.1 \end{bmatrix} \quad D = \begin{bmatrix} d_1 \\ d_2 \end{bmatrix} \text{ and } X = \begin{bmatrix} x_1 \\ x_1 \end{bmatrix}$

are the input-coefficient, final-demand, and output matrices, respectively. We have:

$$I_2 - A \begin{bmatrix} 0.8 & -0.3 \\ -0.3 & 0.9 \end{bmatrix}, \quad (I_2 - A)^{-1} = \begin{bmatrix} \dfrac{90}{63} & \dfrac{30}{63} \\ \dfrac{30}{63} & \dfrac{80}{63} \end{bmatrix}$$

$$X = (I_2 - A)^{-1} D = \begin{bmatrix} \dfrac{90}{30} d_1 + \dfrac{30}{63} d_2 \\ \dfrac{30}{63} d_1 + \dfrac{80}{30} d_2 \end{bmatrix}$$

(b) $5100 worth of commodity 1 and $5200 worth of commodity 2; $1020 worth of commodity 1 is needed to produce commodity 1, and $1560 worth of commodity 1 is needed to produce commodity 2; $1530 worth of commodity 2 is needed to produce commodity 1, and $520 worth of commodity 2 is needed to produce commodity 2.

(c) $6750 worth of commodity 1 and $6450 worth of commodity 2; $1350 worth of commodity 1 is needed to produce commodity

1, and \$1935 worth of commodity 1 is needed to produce commodity 2; \$2025 worth of commodity 2 is needed to produce commodity 1, and \$645 worth of commodity 2 is needed to produce commodity 2.

19.14   Chapter 14 (p. 295)

1. The function $f(t) = N_o e^{-0.0001209t}$ describes the amount of carbon-14 present in the organic remains, $f(t)$, after t years. Since the artifacts found contain 60% of the original amount of carbon-14, we have $f(t) = 0.6N_o$, and thus $0.6N_o = N_o e^{-0.0001209t}$, $0.6 = e^{-0.0001209t}$, $\ln 0.6 = -0.0001209t$
    Thus

$$t = \frac{\ln 0.6}{-0.0001209} = \frac{-0.5108}{-0.0001209} = 4225$$

The artifacts are approximatyely 4225 years old.
Review assumptions 1, 2, and 3.

2. $f(t) = N_o e^{kt}$; from $N_o/2 = 1600$ we obtain $f(t) = N_o e^{-0.0004t}$ where t is time in years and N_o is the amount of radium (in grams) initially present in the given portion of matter.

3. $f(t) = N_o e^{-0.0001209t}$; $0.80N_o = N_o e^{-0.0001209t}$, $\ln 0.8 = -0.0001209t$,

$$t = \frac{0.22314}{-0.0001209} = 1846 \text{ years}$$

19.15   Chapter 15 (p. 305)

Rasa's Experience vs. Andy's Model

1. Rasa found Andy's theorem concerning the travel time from Brooklyn to Kennebunkport on a holiday weekend false. Andy's model did not

realistically take into account traffic delays at the tollgates of the Whitestone Bridge that are characteristic of holiday weekend travel. Rasa's modification of Andy's model was more successful in taking into account this aspect of reality.

2.   **Rasa' s Experience vs. The Behavior of Mercury and Light**

Rasa's vacation trip time (7 hours) proved that Andy's Theorem on the trip time (6 hours) is false concerning the trip she made on the Labor Day weekend. This prompted Rasa to modify Andy's model (see Ch.2). Her model was more successful in taking into allout traffic delay around the Whitestone bridge.

Mercury and Light

It was found that Newton's theorems concerning the behavior of Mercury's orbit and the behavior of light when it passes close to a star were not realistic. Einstein's modification of Newton's model (General Theory of Relativity) was more successful in taking into account this aspect of reality.

3.   "No experiment can ever prove me right" in the sense of establishing that my theory has been proved unequivocally true. "A single experiment may at any time prove me wrong" in the sense that if it establishes that a prediction (theorem) of my theory is off reality's mark, this establishes that my theory needs to be refined.

4.   (a)   There is no debate about the validity of the conclusions of Einstein's general theory of relativity; they follow from the hypothesis of Einstein's theory and this establishes their validity in terms of this theory. The experimental technique cited is irrelevant to this issue.

      (b)   The technique cited might help to resolve questions about the realism of Einstein's conclusion.

7.   It's a question of what one means by confidence. If confidence is understood to mean that a theory of science contains the final truth on the matter in question, then the discovery that it does not can be most unsettling. A scientific theory, but its nature, cannot be considered the

final word on the phenomenon under consideration. It is a model which, sooner or later we can expect, will require refinement as facts that deviate from its predictions (theorems) are uncovered. As facts come to light that agree with a theory's predictions, our confidence in the picture that the theory gives us is strengthened, but it would be a fundamental misunderstanding to view that picture as the final word.

19.16   Chapter 16 (p. 319)

1. Disagree. The parallel postulate problem was to either deduce Euclid's fifth postulate from his other postulates or replace it with an equivalent postulate which satisfied the criteria of being simple and self-evident.

2. Disagree. Lobachevsky sought to show that what is now termed the non-Euclidean Geometry of Lobachevsky, Gauss, and Bolyai is as consistent as other branches of mathematics.

3. No; Euclidean Geometry is not the only possible way of describing space.

4. Disagree. Euclidean Geometry is not synonymous with the term physical space. It is one possible model of physical space, so that if a system contains statements that contradict those of Euclidean Geometry this does not mean that it stands in contradiction to the nature of physical space and therefore cannot be a realistic description of physical space.

5. Disagree. The statements confuses truth with validity.

6. Disagree. Replace consistent by inconsistent

7. No. if there were a "perfect" description of physical space, the equivalent of physical space itself, then, **assuming** that physical space is consistent, the argument would hold up. But there is no such "perfect" description of physical space.

8. Disagree. Proof in geometry (and mathematics more generally) is employed to establish the validity of theorems.

9. Agree. Considering the impact that the discovery of non-Euclidean Geometry has had on mathematics and its applications one must wholeheartedly agree with Hilbert's observation.

19.17   Chapter 17 (p. 324)

1. No; the same sample of people is used for both sets of data.

2. Yes; the day and evening students have little or nothing to do with each other and form, it is reasonable to **assume,** two independent groups.

3. No; the fact that once a spouse is chosen at random the other spouse must be chosen means that the sample is not randomly and independently drawn.

4. Yes; the samples from the two plants were randomly and independently chosen.

20

The Strings Attached to a Math Structure:
A Big Deal?

20.1   Preface

A math structure may be likened to a pair of shoes and the situation to which it is to be applied to feet. If the feet are a "little" smaller or larger than the shoe size, the shoes may still be usable. If they are "significantly" larger than the shoe size the shoes will be very uncomfortable and unusable, no matter how attractive and stylish they may be. With shoes vs. feet it's obvious whether or not the fit is satisfactory.

With math structure vs. situation to which it is to be applied, it's often far from obvious. I invite you to consider the following situations.

20.2   Linearity

Linearity is the first, middle, and last name of linear programming, and chapters 8 and 9 take up a number of situations that were realistically modeled based on the assumption of linearity.

But that's only part of the story. For proper perspective on the linearity assumption (and more generally with all assumptions) I believe that it's to a student's advantage to see examples that illustrate situations where high hopes that were initially placed on linearity were subsequently dashed by confrontation with reality. The advertising-media selection problem provides us with a revealing case study.

Advertising Media Selection

To advertise its new best seller, the Rasa Publishing Company is planning to buy morning and afternoon time on radio station WQRX. Morning time costs $1000 per minute and afternoon time costs $800 per minute. It is estimated (**assumed**) if you prefer, that morning commercials reach 0.9 million listeners and that afternoon commercials reach 0.6 million listeners. At most, 16 minutes of morning time is available in the month in which the advertising campaign is to run. The advertising department of the Rasa Company feels that at least 8 minutes of morning time and at least 6 minutes of afternoon time should be purchased. The advertising budget for this campaign is $24,000.

How much morning and afternoon time should be purchased so as to maximize the total number of listeners reached in the month in which the advertising campaign is to run?

Let x denote the number of minutes of morning time and y the number of minutes of afternoon time to be purchased. The **estimated** number of listeners reached (in millions) is expressed by:

$$F(x, y) = 0.9x + 0.6y$$

The inequalities

$$x \geq 8, \ y \geq 6$$

express the requirement that at least 8 minutes of morning time and 6 minutes of afternoon time are to be purchased. That at most 16 minutes of morning time is available is expressed by:

$$x \leq 16$$

$1000x + 800y$ expresses the cost of x minutes of morning time and y minutes of afternoon time, and since this cost cannot exceed $24,000, we have:

$$1000x + 800y \leq 24,000$$

Therefore, we obtain the following linear program:

$$\text{Maximize } F(x, y) = 0.9x + 0.6y$$

subject to

$$x \geq 0$$
$$y \geq 0$$
$$x \geq 8$$
$$y \geq 6$$
$$x \leq 16$$
$$1000x + 800y \leq 24{,}000$$

This is an example of a general class of problems called **advertising media-selection** problems. The general advertising media-selection problem is to choose from various media capable of carrying an advertisement a selection that is, in some sense, best. Specific choices within a given medium as well as given media are included in the alternatives. The constraints in media selection include the size of the advertising budget, the minimum and maximum usages of specific media categories, and the desired minimum exposure rate to envisioned buyers.

A number of approaches have been developed for the media-selection problem [1], and in the early 1960's hopes ran high in the world of advertising for the use of linear programming. An early linear-programming model for media selection was the one developed by James Engel and Martin Warshaw [2] for the McGraw-Edison Company. The Pennsylvania Transformer division of McGraw-Edison manufactures transformers for use by industrial plants, schools, hospitals, commercial construction projects, and so on. Ten trade publications were considered for advertising purposes, and $25,000 was allocated for industrial advertising for a period of 1 year. Since the purchase decision is usually made by the plant engineer, the objective posed was to maximize the number of plant engineers reached. The following linear program model was developed:

$$\text{Maximize } f = 0X_1 + 15.15X_2 + 32.87X_3 + 49X_4 + 56.65X_5 + 17.54X_6$$
$$+ 58.20X_7 + 0X_8 + 23.53X_9 + 40.00X_{10}$$

subject to nonnegativity of the variables ($X_1 \geq 0, X_2 \geq 0$, etc.)

$$X_1 \leq 5.400$$
$$X_2 \leq 9{,}504$$
$$X_3 \leq 8{,}760$$

$$X_4 \le 10,680$$
$$X_5 \le 11,016$$
$$X_6 \le 5,472$$
$$X_7 \le 9,072$$
$$X_8 \le 3,300$$
$$X_9 \le 8,160$$
$$X_{10} \le 6,900$$
$$X_1 + X_2 + X_3 + X_4 + X_5 + X_6 + X_7 + X_8 + X_9 + X_{10} \le 25,000$$

where X_1 is the amount to be invested in media 1 (*Consulting Engineer Magazine*), X_2 is the amount to be invested in media 2 (*Electrical Construction Magazine*), and so on. The coefficients 0 of X_1, 15.15 of X_2, 32.87 of X_3, and so on, in the linear function f represent the number of plant engineers reached by each magazine per advertising dollar invested, so that f represents the total number of plant engineers reached. The last constraint expresses the condition that no more than \$25,000 is to be spent on advertising and the other constraints are to prevent more dollars from being invested in any one monthly magazine than is necessary to buy 12 insertions.

Although linear program models were satisfactory for crude versions of the media-selection problem, it soon became clear that the features exhibited by more sophisticated versions of the problem could not realistically be modeled in linear program models terms.

Frank Bass and Ronald Lonsdale [3] found linear program models to be

> crude devices to apply to the media-selection problem. The **linearity assumption** itself, is the source of much of the difficulty. Justifying an **assumption** of linear response to advertising exposures on theoretical grounds would be difficult. **Assumptions** about the nature of response to advertising cause most difficulties in models of the type examined in this article.

Philip Kotler [4; p. 478] noted the following limitations:

Linear programming **assumes** that repeat exposures have a constant marginal effect.

It **assumes** constant media costs (no discounts).

It cannot handle the problem of audience duplication.

It says nothing about when ads should be scheduled.

The Rasa Company's linear program model reflects the conditions formulated by its marketing department, but in view of the reservations expressed by Bass, Lonsdale, and Kotler, the conditions formulated and the linear program model developed to express them must be considered a crude version of the advertising media-selection problem.

Although later linear programming approaches to the media-selection problem sought to overcome the criticisms that had been voiced, the message was clear: although linearity, as a mathematical tool ([5] and [6]), is too good not to be true, a linear program model is **not always** a suitable fit for a media-selection problem; that is, the **assumptions** that must be made to force a fit are not always sufficiently realistic. When the model doesn't fit, don't use it.

A Food for Thought Question

1. The automobile manufacturer Chuck Associates plans to advertise its new Chuck IV sports car model on cable station ACB. Late afternoon time costs $10,000 per minute and early evening time costs $20,000 per minute. Late afternoon commercials reach an **assumed** 1.5 million viewers and early evening commercials reach an **assumed** 2.3 million viewers. The marketing department of Chuck Associates believes that at least 12 minutes of late afternoon time and at least 10 minutes of early evening time are needed to effectively communicate the Company's message. At most 20 minutes of early afternoon time is available for the time period that Chuck Associates plans to run its advertising campaign. The budget for the campaign has been set at $100,000.

    The problem is to determine the number of minutes of late afternoon and early evening time to maximize the number of viewers reached.

    (a)   Set up a linear program model for Chuck Associates's problem.

    (b)   In terms of the views expressed by Bass, Lonsdale, and Kotler, should consideration be given to not using this linear program model? Explain.

As an aside, it is noteworthy that the Chuck Associates linear program model has no solution. It provides a simple vehicle for those who wish to pursue this dimension.

20.3   Probability

Random Sampling

At the heart of equally-likely outcome probability modeling is random sampling. When we say that a sample is to be **chosen at random** we have in mind the idea that there is no bias, deliberate or inadvertent, which favors certain samples being chosen over others; a level playing field, so-to-speak.

It's not enough to simply say in class, take a sample at random or a sample is chosen at random, and leave it at that.

For students to obtain an understanding of what these innocuous sounding directions entail it is important that difficulties involved in achieving random sampling in practice be discussed. The issue is pertinent not only to probability but to statistics and polling.

Achieving Random Sampling in Practice.

To raise money for Huxley College's scholarship fund the student Social Science Society sold tickets at the College's homecoming affair. The prize was a 32-inch television set. The tickets were placed in a bowl and mixed. About midway through the festivities Huxley's President, blind folded, reached into the bowl and drew the winning ticket. What everyone expected and **assumed** was that the drawing was fair in the sense that there was no bias in the drawing which favored some tickets being drawn over others—that is, that it was drawn at random.

On the face of it the procedure seems reasonable enough for the task at hand, but how close does it come to satisfying the requirements of a random drawing? The problem is with the physical stirring of the tickets to achieve a "thorough mix." Obtaining a "thorough mix" becomes more and more difficult to achieve as the number of tickets increases, and it is not clear whether early, middle, or late ticket buyers might be favored and to what extent. Still, for the task at hand the degree of randomness achieved might be random enough, except possibly for those individuals who are willing to go to war over a TV set or what they consider the "principle of the matter."

The stakes are considerably higher when big money is involved.

Can We Trust TV Ratings?

The life span of a television program is determined by the public's reaction to it, which is measured by TV ratings. These ratings, produced by the Nielsen Company, estimate the audience in terms of the percentage of those sets in use which are turned to each channel, called a share, or in terms of the percentage of the total possible audience, sets on or off, called a rating. Shares and ratings are further broken down according to the sex and age of viewers so that advertisers can better focus their advertising campaigns. These numbers determine the buying and selling of billions of dollars of television air time. They mean life or death to television programs. The half-hour comedy *Good & Evil*, which had promising ingredients in terms of writing, acting and production talent, had a short life after its premiere in the fall of 1991 because of low initial ratings. In March 1992 NBC announced that they were dropping two successful shows, *Matlock* and *In the Heat of the Night*, because the demographic numbers favored older viewers while the network wished to build around a more youthful audience.

Since 1986 the data which underlie the ratings have been collected by a device called a people-meter. The remote control part of a people-meter rests on top of the television set. When the set is turned on, the meter prompts viewers to enter their identification number. Information is provided on what channels are being beamed into the household and who is watching them. Nielsen puts its people-meter into 4000 households selected at random—that is, without bias—from the approximately 93 million homes in America with television.

The people-meter data gathering system produced lower ratings for the networks than had been expected and a serious question arose as to whether this was because of the increased or decreased accuracy of this system over the method it replaced. The networks commissioned a study of the Nielsen methodology and two years later this Committee on Nationwide Television Audience measurement (CONTAM) issued a nine-volume report that was highly critical of the Nielsen system. The report found evidence of button fatigue—that over time people did not push the buttons that would insure data accuracy as they did in the beginning. CONTAM was highly critical of Nielsen's sampling procedures for obtaining the 4,000 households that make up their sample; random sampling was envisioned in the methodology, but the actual sampling **deviated significantly** from this requirement. From this came ratings which were highly suspect. David Poltrack, senior vice president of research at CBS, observes that: "The whole business is crazy. I don't think there's an advertising agency in the United States that could get up in front of its clients and justify the way business is done right now. It's being bought on narrow based demographics, demographic targets which are not representative of product consumption in the United States." [9]

> Random sampling was called for in theory, but not delivered in practice. This yielded highly suspect ratings.

Nielsen overcame the statistical sampling problem, but it was still plagued by the problem of getting "honest" data from viewers in the sample selected. Its people-meter system for eliciting viewing data has been described as too mechanical and as not being user friendly.

Going to War

When it comes to going to war based on the outcome of a random drawing, the stakes are raised still higher. The European phase of the Second World War and Japan's aggressiveness in Asia greatly alarmed the United States, and on September 6, 1940 Congress passed a conscription law, America's first peacetime draft. To implement the draft each eligible man in a Selective Service District was assigned a number which was put into a capsule. The capsules were put into a bowl, stirred, and then drawn one at a time. The order of the drawing determined the order of the draftees. The properties of the resulting sequence prompted questions about the randomness of the drawing, which gets back to the thoroughness of the mixing (see, for example, [11]).

In 1969 the administration of the draft in the United States to determine the order in which men born in 1950 would be drafted was changed to a lottery system. Three hundred sixty six capsules were prepared (for a leap year), each containing a birthdate. Each month's capsules were put into a separate box. The boxes were emptied into a drum, first those for January, followed by those for February, and so on for the subsequent months. The drum was rotated a few times, the capsules were poured into a bowl, and on December 1, 1969 the drawing was made. Those with birthdays on the capsules drawn first would be drafted first, and so on. If you birthday fell among those drawn last, there was a good chance that you would not be drafted at all. The results of the drawing are given in Table 20.1, from which we see that the earlier months, January through June got the larger share of the last-to-be-drafted numbers and the later months, July through December, got the larger share of the first-to-be drafted numbers.

Table 20.1

Month	1-22 (First-Drafted)	123-244 (Middle)	245-366 (Last Drafted)
January	9	12	10
February	7	12	10
March	5	10	16
April	8	8	14
May	9	7	15
June	11	7	12
July	12	7	12
August	13	7	11
September	10	15	5
October	9	15	7
November	12	12	6
December	17	10	4

The results suggest the possibility that the earlier months' capsules were concentrated at the bottom of the bowl, while those of the later months were concentrated at the top and were more accessible for picking. Formal hypothesis tests of randomness did not support the hypothesis that a random drawing had been carried out (see, for example, [8] and [10]).

This period was one of great turbulence in American history and slogans such as "Draft Beer, not Students" and "Hell No, We Won't Go," were a prominent part of the scene. Perceptions of an unfair draft lottery on top of what was considered by many to be an indefensible draft for an indefensible war added fuel to a raging fire. In response to criticism the Selective Service modified its number selection mechanism for the draft lottery conducted in 1970. What these examples serve to make clear, however, is the difficulty of achieving a random selection in practice.

Another way of obtaining a random sample which avoids the difficulties of obtaining a sufficiently thorough mix of tickets or capsules is by use of a table of random numbers.

Random Numbers vs. Pseudorandom Numbers

The study of many complex phenomena requires the generation of large streams of random numbers. It came as quite a shock when three scientists showed that five of the most often used computer programs for generating random numbers induced errors in the study of the behavior of atoms in a magnetic crystal because the numbers produced were not random, despite the fact that they passed several statistical tests for randomness [7]. The deviations from randomness were subtle and, although the pseudorandom numbers produced were satisfactory for many purposes, they were not satisfactory for the problem at hand.

Is it possible that no machine based system can produce truly random numbers? John von Neumann, regarded as the father of the modern computer, thought that the answer is yes. In an observation made in 1951 von Neumann expressed the view that anyone who believed a computer could produce truly random numbers was living is a state of sin. It may be that the best we can hope to do is produce pseudorandom numbers which are satisfactory for the purpose at hand, and that the truly random number is a mathematical ideal which cannot be attained. The question that arises in an application situation then is, how close to random is random enough?

Food for Thought Questions

2. To carry out a marketing survey on consumer preferences for kitchen appliances Elias Marketing Research Associates placed two interviewers on the busiest street in town to interview passersby. Does the sample of opinions obtained qualify as a random sample? Explain.

3. Tickets were sold at Ecap University's graduation celebration to help raise funds for the University's new library. The tickets sold were placed in a bowl as soon as they were sold. At the end of the graduation festivities the University's Library Director, Harriet Warren, reached into the bowl and choose a ticket at random. The ticket holder was awarded a newly published edition of Charles Dickens's collected works.

   Kevin Reynolds, who was among the first to purchase a ticket, protested that the drawing procedure was biased and demanded that ticket purchasers be given a refund or that the drawing be held again. "Your claim is not justified Mr. Reynolds," replied the Dean of Student Affairs. "Ms. Warren was blindfolded and the ticket was chosen at random."

   Who is right? Explain.

Markov for Marketing?

For perspective on an application of a mathematical structure I believe it's important for students to see examples illustrating situations where high hopes placed on the mathematical structure did not pan out because of confrontation with reality.

For probability a revealing example is provided by the study of brand-choice behavior in marketing and attempts to enlist stationary Markov, chains in its study. To discuss this example in class I focus on the modeling dimension. It's not necessary to get into specifics about Markov chains and matrix algebra.

The late 1950s and 1960s saw the development of Markov chain models to describe consumer brand choice behavior. A starting point for many of these investigations is the view that a brand loyalty and brand switching matrix of probabilities can be constructed from data on sequences of consumer purchases.

$$
T_1 = \begin{bmatrix}
p_{11} & p_{12} \cdots p_{1n} \\
p_{21} & p_{22} \quad p_{2n} \\
\vdots & \vdots \quad \vdots \\
p_{n1} & p_{n2} \cdots p_{nn}
\end{bmatrix}
$$

The value p_{11}, for example, express the probability that the consumer, having bought brand 1 in the last period, will also purchase brand 1 in the next period. More generally, p_{ij} expresses the probability that the consumer, having bought brand i in the last period, will purchase brand j in the next period.

An early application of this sort is one undertaken by Benjamin Lipstein [19] concerning the test marketing of a new margarine, fictitiously called Electra, in the Chicago area from November 1958 to May 1959. In Lipstein's study the possible states a margarine buyer could be in were the following

E_1:   Electra Brand
E_2:   Gloria
E_3:   B-R Stores brand
E_4:   Aunt Mary's brand
E_5:   Meadowlark brand
E_6:   All other brands
E_7:   Did not buy margarine during time period

Lipstein's paper contains the brand loyalty and brand switching matrix of transition probabilities (Table 20.2) which represents the situation in the margarine market in Chicago shortly after the introduction of the new brand Electra.

Table 20.2

Next period \ Original Period	Electra	Gloria	B-R	Aunt Mary's	Meadow-Lark	Other	Did not buy
Electra	.12	.05	.03	.02	.04	.03	.05
Gloria	.05	.25	.02	.05	.01	.05	.03
B-R	.07	.03	.21	.01	.03	.03	.04
Aunt Mary's	.04	.02	.05	.23	.02	.04	.01
Meadowlark	.03	.02	.03	.04	.22	.05	.02
Other	.28	.26	.26	.25	.30	.23	.28
Did not buy	.41	.37	.40	.40	.38	.57	.57

Beware the Assumptions

But are Markov chain models realistic for the study of consumer brand choice behavior? A critical appraisal was given by A.S.C. Ehrenberg [12], who expressed the view that "frequent public reference to Markov-brand switching models had not been matched by an obvious array of published demonstration of their practical effectiveness." On the basis of a detailed discussion and analysis, Ehrenberg concluded that

> the failure of the Markov brand-switching model to live up to its
> earlier public reputation need not be surprising if seen as an example

of misguided but perhaps understandable enthusiasm for forcing an attractively simple piece of college mathematics (stationary Markov theory) onto repeat-buying and brand-switching data while:

1. Omitting to ensure that the data are of a technically suitable form to be modeled by the model.

2. Omitting to examine the crucial **assumption** involved.

3. Omitting any self-critical appraisal of the various concepts and analytical steps in the approach.

4. Omitting to gather any generalized empirical knowledge of repeat-buying and brand-switching behavior as such.

William F. Massey and Donald G. Morrison [23] expressed agreement with many of Ehrenberg's arguments that the simple Markov chain does not fit all, or even many, real brand-switching situations, but felt that Ehrenberg had been too harsh in his judgment. They expressed the view that the basic Markovian approach is fruitful and should not be abandoned. In a reply [13], Ehrenberg took issue with Massey and Morrison and again raised the question, "Can we not bury Markov for Marketing?"

The Ehrenberg-Massey-Morrison exchange should serve to remind us that mathematical models can only provide us with **valid conclusions with respect to our assumptions**.

If the assumptions are unrealistic, then the mathematical model does not properly fit the situation, and to try to force a fit can be as counterproductive and painful as forcing a pair of size 8 shoes on feet that require size 10.

Ehrenberg Triumphant

As to Lipstein's study, it follows from his transition matrix that over the long term Electra brand would end up with about 4% of the margarine market. However, six months or so after its introduction. Electra succeeded in capturing about 12% of the market. Electra had been effective in building up the percentage of buyers who having purchased Electra in one period, remained loyal to it in the next period, so that the 0.12 value in the first row, first column of Table 20.2 would have to be revised to 0.23.

The need to change transition probabilities had not been taken into account in Lipstein's model.

20.4   Statistics

Random Sampling

Since the **equally-likely outcome assumption** is a basic underpinning of statistics the problem of achieving random sampling in practice is shared by a wide spectrum of statistical applications.

Robustness

Hypothesis tests require that **conditions** of one sort or another be satisfied. Some of these strings attached, such as the normality of the population(s) involved and equality of population variances, are rather severe. An important question concerning applications is, to what extent can we relax the required strings attached and still apply the test? A test that remains reliable under strong modifications of the strings attached on which it is based is said to be **robust**.

The following is a revealing example which illustrates that robustness is not a condition that we can take for granted.

The test employing a chi-square variable to test the hypothesis $H_o : \sigma^2 = \sigma^2_o$ is an example of a test that lacks robustness with respect to the **assumed** normality of the underlying populations. Hoel, Port, and Stone [25] take up the case where the underlying population is distributed according to $y = xe^{-x}$, $x > 0$ (see Figure 20.1). they show that the accept/reserve judgment

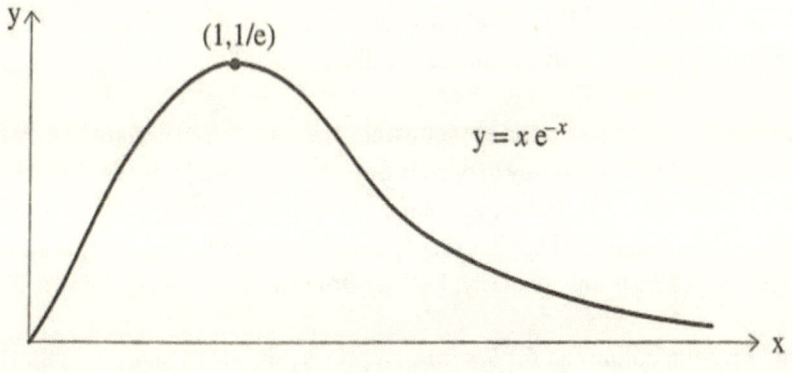

Figure 20.1

interval is approximately 1.58 times as long as the one obtained under the **incorrect assumption** of normality, so that the null hypothesis $H_o : \sigma^2 = \sigma^2_o$, versus the two-sided alternative, would be rejected much more often than need be.

What do we do when we are confronted by an application situation where the conditions required for the use of the appropriate test are not satisfied or it not clear whether they are "reasonably well" satisfied? We turn to non-parametric tests.

20.5   Economic Assumptions

Let's Assume That . . .

> This directive underlies all applications of mathematics to real world situations. It's far too often the case that no sooner is the directive stated, its reality is ignored.

If I had to make a list of the most numerous practicioners of this practice, economists would receive prominent billing.

An engineer, a chemist, and an economist are marooned on an island and they find a can of tuna, but have no way to open it. Says the engineer: "Let's find a rock and crush the can open." Suggests the chemist: "That's impractical: A better way is to find some chemicals and blast it open." Quoth the economist: "You people are truly misguided. There is only one way: Let's **assume** we have a can opener . . . ."

Gene Epstein, columnist for *Barron's*, reprises this tale to make an important point about commonly made **economic assumptions**. Specifically, Epstein considers the article "Labour Allocation in a Cooperative Enterprise" by Amartya Sen, who a week earlier had been announced as the 1998 winner of the Nobel Prize for Economics. Epstein notes that Sen sets down **assumptions** which, he assures us, are "serious but not especially odd in this branch of economics." They include "well-behaved utility and production functions; automatic fulfillment of the second order conditions of welfare maximization and of equilibrium; no uncertainty; perfectly competitive markets; homogeneity of labor . . .—a whole set of can openers" whose realism, Epstein points out, is "truly dizzying for the colossal naivete they reveal." More generally, "Your typical economist," Epstein further observes, "**assumes** such absurdities as 'no uncertainty' in markets, even though uncertainty is any market's middle name, and 'well-behaved utility and production functions' (read: unchanging consumer tastes and production processes), even though such good behavior is the rare exception rather than the rule—and he **assumes** so for only one reason: Because that way, **he gets to use a lot of math**" [26].

Epstein, as most of us, is not opposed to using a lot of math provided that the math conditions are suitable fits to the economic situation.

In the essay "The Future of Economics," *The Economist* puts it cogently: If models are to reveal anything, they muset be simpler than reality: the challenge is to simplify usefully" [27]. Also see [28] and [29].

20.6   Deviation in Practice from the Requirements of Theory

As we appreciate from our daily experience, deviations from a stated norm may have consequences ranging from negligible to catastrophic. If a "little more or less" of a delicate herb than specified is added in the preparation of a gourmet meal, for example, the effect would in all likelihood be inconsequential. If we are off the norm by more than a "little," a gourmet delicacy could easily be turned into a gourmet disaster. Medication that is life saving when taken as specified might become life threatening when the prescribed dose is exceeded.

Deviation from mathematical requirements may, in a similar vein, have negligible or profound consequences, depending on the situation.

> The profound lesson to be gleaned from these considerations is that when conditions are stipulated, whether they be in connection with sampling or of a more general nature, it is our obligation to insure, as best we can, that they are met if the results obtained from practice are to be in close agreement with the conclusions derived from theory.

20.7   Food for Thought Questions

4. (a)   What do we mean when we say that a hypothesis test is robust?

    (b)   Why is the robustness of a test an important issue?

20.8   References of Interest?

For 20.2

1. D. Genech, "Different Approaches to Advertising Media Selection," *Operational Research Quarterly,* vol. 21, no. 2 (June 1970), pp. 193-219; Philip Kotler, *Marketing Management* (Englewood Cliffs, N.J.: Prentice-Hall, Inc., 1967), Chapter 18.

2. J. Engel and M. Warshaw, "Allocating Advertising Dollars by Linear Programming", *Journal of Advertising Research*, vol. 4, no. 3 (Sept. 1964), pp. 42-48.

3. F. Bass and R. Lonsdale, "An Exploration of Linear Programming in Media Selection", *Journal of Marketing Research*, vol III, no. 2 (May 1966), pp. 179-188.

4. P. Kotler, *Marketing Management* (Englewood Cliffs, N.J.: Prentice-Hall, Inc., (1967).

5. A. Charnes, W.W. Cooper, J.K. DeVoe, D.B. Learner, and W. Reinecke, "L.P. II: A Goal Programming Model for Planning," *Management Science*, vol. 14, no. 8 (April 1968), pp. 423-430.

6. A Charnes, W.W. Cooper, D.B. Learner, and E.F. Snow, "Note on an Application of a Goal Programming Model for Media Planning", *Management Science*, vol. 14, no. 8 (April 1968), pp. 431-436.

For 20.3

7. M. Browne, "Coin-Tossing Computers Found to Show Subtle Bias," *The New York Times*, Jan. 12, 1993, p. c1.

8. C. Hawkins, J. Weber, *Statistical Analysis: Applications to Business and Economics* (Harper and Row, 1980), 297-303.

9. Nova, "Can You Believe TV Ratings?" WGBA, Boston, 1992.

10. J. Rosenblatt, J. Filliben, "Randomization and the Draft Lottery," *Science,* 171 (1971), 306-308.

11. S. Stouffer, W. Bartky, statement, *Chicago Tribune*, Nov. 2, 1940, p. 4.

For 20.4

12. A.S.C. Ehrenberg. "An Appraisal of Markov Brand-Switching Models," *Journal of Marketing Research*, vol. 2, no. 4 (Nov. 1964), pp. 347-362.

13. _____. "On Clarifying M and M," *Journal of Marketing Research*, vol. 5, no. 2 (May 1968), pp. 228-29.

14. Jean E. Draper and Larry H. Nolan. "A Markov Chain Analysis of Brand Preference." *Journal of Advertising Research*, vol. 4, no. 3 (September 1964), pp. 33-39.

15. Frank Harary and Benjamin Lipstein. "The Dynamics of Brand Loyalty: A Markovian Approach." *Operations Research*, vol. 10., no. 1 (January-February 1962), pp. 19-40.

16. Jerome D. Herniter and John F. Mag. "Customer Behavior as a Markov Process." *Operations Research*, vol. 9, no. 1 (January-February 1961), pp. 105-122.

17. Ronald A. Howard. "Stochastic Process Models of Consumer Behavior." *Journal of Advertising Research*, vol. 3, no. 3 (September 1963), pp.

35-42. Reprinted in *Marketing Models: Quantitative Applications*, edited by R. L. Day and L.J. Parsons, pp. 104-117. Scranton, Pa.: Intext Educational Publishers, 1971.

18. Benjamin Lipstein. "The Dynamics of Brand Loyalty and Brand Switching." *Proceedings of the Fifth Annual Conference of the Advertising Research Foundation* (1959), pp. 101-108.

19. _____. "Tests for Test Marketing," *Harvard Business Review*, vol. 76 (March-April, 1961), pp. 365-369.

20. _____. "A Mathematical Model of Consumer Behavior." *Journal of Marketing Researchi*, vol. 2, no. 3 (August 1965), pp. 259-265. Reprinted in *Marketing Models: Quantitative Applications*, edited by R. L. Day and L. J. Parsons, pp. 65-79.

21. P.A. Longton and B. T. Warner. "A Mathematical Model for Marketing." *Metra*, vol. 1 (September 1962), pp. 297-310.

22. Richard B. Maffei. "Brand Preferences and Simple Markov Processes." *Operations Research*, vol. 8, no. 2 (March-April 1960), pp. 210-218.

23. William Massey and Donald Morrison. "Comments on Ehrenberg's Appraisal of Brand-Switching Models." *Journal of Marketing Research*, vol. 5, no. 2 (May 1968), pp. 225-227.

24. George P. H. Styan and Harry Smith Jr. "Markov Chains Applied to Marketing." *Journal of Marketing Research*, vol. 1, no. 1 (February 1964), pp. 50-55.

For 20.5

25. P. Hoel, S. Port, C. Stone; *Introduction to Statistical Theory* (Boston: Houghton Mifflin, 1971), pp. 182-84.

For 20.6

26. G. Epstein, "Is it Really Reasonable to Assume that the Newest Nobelist Deserved the Prize?" *Barron's*, Oct. 19, 1998.

27. *The Economist*, "The Future of Economies", March 4, 2000

28. M. Weinstein, "Students Seek Some Reality Amid the Math of Economics," *The New York Times*, Sept. 18, 1999.

29. J. Galbraith, "How the Economists Got It Wrong", *The American Prospect*, Feb. 14, 2000.

20.9   Comments on Selected Food for Thought Questions: 2

1. (a)   **(p. 372)** Max. $L(x, y) = 1.5x + 2.3y$

   where x is the number of minutes of late afternoon time to be purchased, and y is the number of minutes of early evening time to be purchased.

   subject to

   $x \geq 0, y \geq 0$

   $x \geq 12$

   $y \geq 10$

   $y \leq 20$

   $x + 2y \leq 10,$

   (b)   By Bass, Lonsdale, and Kotler standards it is a crude linear program model. Whether or not it is sufficient for the needs of Chuck Associates its management will have to decide.

2. **(p. 378)**

   No. The make-up of the busiest street is not necessarily (and probability not) the same as the make-up of all streets in the city.

3. If the tickets were thoroughly mixed so that it is **realistic to assume** that each ticket in the bowl had the same chance of being selected, then the dean is right. However, if this is not the case, then Reynolds is correct.

4. **(p. 386)**

    (a)   A robust hypothesis test is one that remains applicable when its prerequisite conditions are not "fully" satisfied.

    (b)   Robustness is an important issue because it allows us to still perform a hypothesis test and get reliable results when its **assumed** conditions are not fully met.

21

Reliability: Are These Numbers Trustworthy?

21.1   Preface

Slippery Numbers or Mathematical Proof?

Recently the Dean of Administrative Affairs at Huxley College expressed to a friend his concern about rumors that his work is counter-productive.

Dean: "Such talk is totally unjustified and unfair."

Friend: "You're quite right. Have you pointed out that since you do no work at all, even in a leap year, your non-efforts could at worst be considered neutral. In fact, by not participating in Administration Council sessions you have made a strong positive contribution to the effectiveness of that organization."

Dean: "This sounds promising. How could I present a convincing argument?"

Friend: "There are 366 days in a leap year. Now you sleep 8 hours a day. Thus, one-third of the day is spent sleeping. You sleep one-third of the year, or 122 days. This leaves 244 days. Four hours a day are spent on rest, recreation and personal business. Thus one-sixth of the day is spent in this manner. This takes up one-sixth of the year, or 61 days, which leaves you 183 days. Of these 183 days you don't work on Saturdays or Sundays. Since there are 104 Saturdays and Sundays, this leaves you 79 days. How long do you take for meals every day?"

Dean: "It varies, but about 2 hours a day."

Friend: "Two hours a day. In a week you spend more than one-half a day eating. About 28 days are spent sitting around and eating. Twenty-eight days from 79 leaves 51 days. As a top administrator how much vacation do you get?"

Dean: "Seven weeks."

Friend: "That's 49 days from 51. This leaves 2 days; New Year's and Labor Day, which you get off. You don't even work one day in the year and your non-existent work has in fact left Huxley a much better place."

Dean: "Thanks. You always come through for me."

This variation on one of Fred Allen's routines suggests that numbers can be employed in almost any situation, sometimes with hilarious results. *

Alas, many of the slippery number routines we come into contact with in modern life sow confusion rather than amusement. I would submit that this dimension should be given a prominent role in own teaching of mathematics (statistics in particular)

The best form of immunity against slippery numbers is conferred by examination of a number of slippery number routines. Should you wish to incorporate this dimension of numbers in your teaching of mathematics, you might find the following illustrative examples illuminating.

---

*   The power of numbers to deceive is truly frightening. I presented the afore scenario to one of my classes a number of years ago and one of my students insisted that I had proved the result because mathematics does not lie.

21.2   Caution! Slipper Numbers Sold Here

Case 1 Are Statistically Dangerous Schools Necessarily Dangerous? Are Statistically Safe Schools Necessarily Safe?

In July 1986 the New York City Board of Education issued a list of its most dangerous schools based on incident and crime reports that it had received. One respondent notes that he never felt unsafe or threatened in teaching at the fifth listed "most dangerous" junior high school. [15] Another respondent comments: "I believe that the scorecard . . . names not the most dangerous schools but the schools whose administrators have the courage to report what is really happening." [12]

In June 1994 New York City Schools Chancellor Ramon Cortines rejected data on violence in the city's schools, saying that he suspected school administrators were underreporting acts of violence to make their schools appear less turbulent. [3].

In September 1995 Edward Costikyan, the chairman of the Mayor's Commission on School Safety, observed: 'There seems to be a total absence of any reliable numbers on anything. How can you manage anything without knowing what you're dealing with.' [17]. Are things better now? Well . . .

In September 2007 New York City comptroller William C. Thompson Jr. stated that an audit showed that the city had not ensured that all principals accurately report violence in their schools, making it difficult for the public to assess their safety. [7]. Are things better now? Well . . .

Case 2 These Data May Give You Nightmares

Halcion, manufactured by the Upjohn Company and introduced in the United States in 1983 is one of the world's best known sleeping pills. Its main advantage over competing products, Upjohn claimed, is in encouraging nighttime sleep without daytime drowsiness.

How safe is Halcion? It received Food and Drug Administration approval and its manufacturer claims that it is just as safe as other drugs of its kind. Dissenters argue that Halcion is more likely to cause symptoms such as amnesia, paranoia, and depression and that Upjohn engaged in data manipulation to conceal its side effects. This view emerged from a law suit filed by Ilo Grundberg, who killed her mother the day before her mother's 83rd birthday and placed a birthday card in her hand. Mrs. Grundberg claimed that Halcion had made her psychotic, and charges against her were eventually dismissed. Upjohn settled the lawsuit with Mrs. Grundberg before it was to go to trial in August 1991, but in preparation for the suit it had to make available a good deal of data about Halcion to the plaintiff's attorneys.

Dr. Ian Oswald, who was head of the department of psychiatry at the University of Edinburgh and had spent 30 years doing research on sleep, was obtained as an expert witness. Dr. Oswald spent two years going over Upjohn's data and concluded that Upjohn had known about the extent of the drug's adverse effects for 20 years and concealed these data. He concluded that "the whole thing had been one long fraud." [11]. Dr. Graham Dukes, former medical director of the Dutch drug regulatory agency, who examined some of Upjohn's data, believed that the data on Halcion had been organized in such a way as to minimize the drug's adverse effects and that this could not have occurred accidentally.

In reaction to the criticisms voiced, Britain, the Netherlands and Belgium were led to remove the sleeping pill from the market. A report issued in April 1994 by F.D.A. investigators stated that the Upjohn Company had engaged in ongoing misconduct with Halcion. The F.D.A. will investigate, it was announced. As far as I know, we have not heard anything further.

Case 3 Top of the Line Deception

In 1992 the General Accounting Office audited seven "Star Wars" tests conducted between 1990 and 1992. It found that four of the test results described to Congress as successes were false whereas the three tests that were described as complete or partial failures were correct. [19].

Case 4 Spin vs. Counterspin

Speaking on television on Tuesday night of 3 August 1993, President Clinton described the budget legislation then before Congress as "the largest deficit reduction in history." Almost immediately after the President spoke, Senator Robert Dole, Republican leader in the Senate, described the legislation as "the largest tax increase in world history."

"Who is right? Neither; when the dollar amounts are adjusted for inflation so that dollar comparisons are meaningful, 1993's budget bill is neither the biggest reduction measure nor the biggest tax increase in recent years. In 1993 dollars, the bill would lower the annual deficit by a projected total of $496 billion over five years; $241 billion of this would come from tax increases. The bill signed by George Bush in 1990 contained $532 billion in deficit reduction in terms of 1993 dollars. The bill signed by Ronald Reagan in 1982 raised taxes by $286 billion over five years in terms of 1993 dollars.

For discussion of how to take inflation into account in comparing dollar amounts in different time periods, see Determining "Real" Dollar Amounts, sec. 10.5.

Case 5 Does This Scenario Add Up?

In the spring of 480 B.C. the Persian King Xerxes launched against Greece the largest military operation the world had seen to that point. But how large? The Greek historian Herodotus gives the fighting force as 2,641,610 men. After considering the service train that went with them, Herodotus put the total force at 5,283,220 men. But, he adds:

> no one could give the exact number of women who baked the
> bread, or of the concubines, or the eunuchs, or the transport animals
> and baggage-carrying cattle and Indian dogs that came with the
> army—of all these creatures no one could count the numbers, they
> were so large. [9]

The invasion force was large, no doubt, but 5,283,220 men plus others? This exceeds by far the entire population of Greece at that time. The logistic problems of maintaining such a force are, to put it mildly, considerable.

Case 6 There's More Here Than Meets The Eye

Prior to the 1992 election, President George Bush and Secretary of Education Lamar Alexander were hard at work promoting education reform based on the voucher system. A G.I. Bill for kids was envisioned as providing $1000 for each of 500,000 children from low and middle income families in fifty communities to help them pay for a private school education. Envisioned cost: $500 million.

If one considers the approximately 5 million students attending private schools with voting parents, friends and supporters who would strongly feel that equity requires that they be granted such relief, or no votes, the cost is brought to a total in the neighborhood of $5.5 billion.

With expectations having been aroused, suppose that a relatively small number of additional families with children currently in public schools say that they want their children to attend private schools with the government paying at least part of the bill. The political pressure to accommodate an additional 1 to 2 million students in the voucher system could be enormous. If successful, this could add an additional $1 billion to $2 billion to the bill.

Case 7 Beware Backing-in

Backing in is the practice of determining the numbers you want and then generating the assumption needed to obtain them.

In his article *" Tweaking Numbers to Meet Goals Comes Back to Haunt Executive"* (The New York Times, June 29, 2002, AZi) Alex Berenson notes:

> Instead of first figuring out their sales and subtracting expenses to calculate the profit, they [some companies] work backward. They start with the profit that investors are expecting and manipulate their sales and expenses to make sure the numbers come out right."

For discussion of backing-in to show projected Social Security deficits see ch. 18 sec. 18.3

Case 8 Slippery Numbers: An International Dimension

The Slippery Numbers Society (SNS) has an international clientele. Here are a few examples.

Brazil

In a frank conversation between television interviews that was inadvertently broadcasted across his country, Brazil's finance minister Rubens Ricupero expressed the sentiments of many kindred spirits when he confessed of economic indicators: "I have no scruples, what is good we take advantage of. What is bad, we hide" [1]. Minister Ricupero was immediately dismissed, but was this because of his performance or indiscretion?

Britain

In the past *The Economist* has been critical of the U.K. Central Statistical Office as having 'figures often tasting of fudge.' [5], [6].

China

Chinese government statistics have run a gamut of slipperiness. After the Communist Party assumed control in 1949, government statistics were systematically distorted to serve the wishes of the new political establishment. During the period of the Cultural Revolution of the late 1960s and early '70s data-gathering was abandoned as unscientific.

Since the passing of the Cultural Revolution, data-gathering and the publication of state statistics has resumed and other pressures have developed. In May 1994 Zhang Sai, director of the State Statistical Bureau, 'warned that distorted statistics are increasing tensions between Beijing and localities. [18]

Foreign investors in China are wary of Chinese statistics and many have taken to generating their own.

Japan

By late July 1998 American financial experts reached the conclusion that the magnitude of Japan's banking crisis was far worse than had been publicly acknowledged. The bad debts were estimated as being on the order of $1 trillion, nearly twice the official estimate. The true amount, financial experts emphasized, is hard to pin down because Japanese banks have been using accounting tricks to conceal debts that are not being paid. [13]

Soviet Union and Russia

From the beginnings of the Soviet State, Soviet statistics have acquired a reputation of being unreliable. (See [2], [14]). Writing in 1990, V.N. Kirichenko, Chairman of the USSR State Committee on Statistics, expressed a hope to 'ensure the accuracy of the data . . . restore the trust in such data on the part of the Soviet and international public. The country can no longer afford to seek the right way with the help of trick mirrors.' [10], [4]

Since the breakup of the Soviet Union, Russia has continued to have problems with government statistics, but for a different reason. Rather than exaggerating output the statistical pendulum has swung to the extreme of underestimating it. In June 1998 Russia's top statisticians were arrested on charges of manipulating data to underestimate the production of Russian businesses to help them minimize their tax obligations [8].

United States

In June 1998 the thrust of the Republican majority in the House of Representatives was to cut taxes beyond what was called for in the earlier balanced-budget agreement. But then there are the spending cuts needed to achieve balance. The Congressional Budget Office did not produce the numbers needed for this to work out, which prompted the Republican leadership to address a letter to the Appropriations subcommittee warning that if the C.B.O. did not begin to produce better numbers, 'we must review [its] structure and funding.' [14].

21.3   Food for Thought Questions

In replying to the questions posed consider the articles cited to establish a context and other relevant information that you might obtain.

1. "These numbers are all extremely fuzzy." How so? E. Schmitt, T. Shanker, "Taliban and Qaeda Death Toll In Mountain Battle Is a Mystery," *The New York Times*, March 14, 2002; Al. B. Bearak, "Afghans Declare Mountain Victory; Foes' Toll Unclear," *The New York Times*, March 13, 2002; A1

2. The math in President George Bush's proposed 2003 budget is fuzzy? How so? "George Bush's Budget: Unfurl the Fuzzy Maths," *The Economist*, Feb. 9, 2002, 27.

3. Are Texas's education numbers too good to be true? P. Schrag, "Too Good To Be True," *The American Prospect*, Jan. 3, 2000; 46-49.

4. Are America's productivity numbers for the late 1990's as good as they once seemed to be? "Measuring the New Economy," "A Spanner in the Productivity Miracle," *The Economist*, Aug. 11, 2001; 12-13, 55-56.

5. How do the numbers concerning the impact of oil drilling in the Arctic National Wildlife Refuge not lie, strictly speaking, but not tell the truth? P. Krugman, "Two Thousand Acres," *The New York Times*, March 1, 2002; A23.

6. What is the problem of getting an "accurate" count of race in the census? R. Thornton, "What the Census Doesn't Count," *The New York Times*, March 23, 2001; A19.

7. Misleading numbers in medical studies? How so? "Watch for Weasel Numbers in Medical Studies," *Business Week*, June 17, 2002; 85.

8. Are data reported from company funded medical studies trustworthy? P. Raeburn, "The Credibility Gap in Drug Research," *Business Week*, June 24, 2002; 28.

9. What is the problem with African Numbers? N. Onishi, "African Numbers, Problems and Number Problems," *The New York Times*, Aug. 18, 2002; Wk-5.

10. How were the tests on a missile interception program rigged? T. Weiner, "Lies and Rigged 'Star Wars' Test Fooled the Kremlin, and Congress," *The New York Times*, Aug. 18, 1993; A1; T. Weiner, "General Details Altered 'Star Wars' Test," *The New York Times*, Aug. 27, 1993; A1; E. Schmitt, "Aspin Disputes Report of 'Star Wars' Rigging," *The New York Times*, Sept. 10, 1993; A1; T. Weiner, "Inquiry Finds 'Star Wars' Tried Plan to Exaggerate Test Results," *The New York Times*, July 23, 1994; A1.

11 Prison, populations skew the census? How so? A. Thompson, "Democracy Behind Bars", *The New York Times*, Aug. 6 '09; A23.

12. A problem with economic statistics? How so? "Damn Lies," *The Economist*, Nov. 23, 1996; 18.

13. Do approved cigarette tests provide reliable measures of tar doses that smokers are subjected to? R. Kerber, "Do Approved Cigarette Tests Understate Tar?" *The Wall Street Journal*, Jan. 30, 1997; B1.

14. Is the sharp drop in crime in recent years real or manufactured? F. Butterfield, "As Crime Falls, Pressure Rises to Alter Data," *The New York Times*, Aug. 3, 1998; A1; A. Fine, "Philadelphia Takes Lead in Cleaning Up Crime Statistics," *The Christian Science Monitor*, Dec. 28, 1998; 2.

15. Tainted testing? S. Kilman, "Validity of Mad-Cow Tests Questioned," *The Wall Street Journal*, Jan. 16, 2004; A3.

16. Did questionable numbers determine the outcome of the 1876 presidential election? What were the consequences? N. Kleinfeld, "President Tilden?

No, but Almost, in Another Vote That Dragged On," *The New York Times*, Nov. 12, 2000; 30. "Watch Yourself at Dinner, Dubya," *The Economist*, Nov. 25, 2000; 30.

17. A case of bi-partisan book-cooking? D. Francis, "Budget Fudge," *The Christian Science Monitor*, Aug. 25, 1999; 8.

18. A strong touch of Enron about it? How so? P. Krugman, "Bush's Aggressive Accounting," *The New York Times*, Feb. 5, 2002; A25.

19. Can a seemingly simple number be anything but straightforward? A Tergesen, "Cash-Flow Hocus-Pocus," *Business Week*, July 15, 2002; 130, 132.

20. What are your options for financial hocus-pocus? P. Krugman, "Flavors of Fraud," *The New York Times*, June 28, 2002; A27.

21. Why do companies report one set of numbers to the IRS and a different set to shareholders, and what does this mean for shareholders? A. Tergesen, "How to Spot Tax Tinkering: See If Products Are Really Boosting Profits," *Business Week*, May 20, 2002; 142, 144.

22. Fraudulent arguments and numbers that may seem superficially plausible? How so? P. Krugman, "Bad Faith Economics", *The New York Times*, Jan. 26, 2009; A23.

23. Is Mayor Bloomberg trying to keep control over our schools by exaggerating success? What do the numbers suggest and how reliable are they/ What other numbers should be looked at? D. Ravitch, "Mayor Bloomberg's Crib Sheet", *The New York Times*, April 10, 2009 A23.

24. Citigroup, after devastating financial losses and three government bailouts, announced that its net profit for the first quarter of 2009 was $1.6 billion. A solid or junk figure? E. Dash, " Sharp Pencil Lets Citigroup Declare Profit," *The New York Times*, April 18, 2009 A1.

25. Is China over stating it's true rate of growth? "Economic Focus: The art of Chinese message", *The Economist,* May 23, 2009; p.82.

21.4   References of Interest?

1. J. Brooke, "In Brazil, Slip of the Tongue Makes Campaign Slip," *The New York Times,* Sept. 5, 1994.

2. C. Clark, *A Critique of Russian Statistics* (London: MacMillan and Co., 1939).

3. S. Dillon, "Report Finds More Violence in the Schools," *The New York Times*, July 7, 1994.

4. J. Duncan, A. Gross, *Statistics for the 21st Century* (Chicago: Irwin, 1995); p. 66

5. *The Economist,* "The Good Statistics Guide," Sept 7, 1991; p. 88.

6. *The Economist,* "The Good Statistics Guide," Sept. 11, 1993; p. 65.

7. E. Gootman, "Undercount of Violence in Schools: Defective Reporting is Found at 10 sites", *The New York Times*, Sept. 20, 2007.

8. M. Gordon, "Moscow Statisticians Accused of Aiding Tax Evasion," *The New York Times*, June 10, 1998.

9. Herodutus, *The History*; translated by D. Grene (Chicago: The Univ. of Chicago Press, 1987), pp. 535-6.

10. V. Kirichenko, "Return Credibility to Statistics," *Business Economics*, Oct. 1990, pp. 50-57.

11. G. Kolata, "Maker of Sleeping Pill Hid Data on Side Effects, Researchers Say," *The New York Times*, Jan. 20, 1992.

12. E. Richman, Letter, *The New York Times*, Aug. 12, 1986.

13. D. Sanger, "Bad Debt Held by Japan's Banks Now Estimated Near $1 Trillion," *The New York Times*, July 30, 1998.

14. H. Shaffer (ed.), *The Soviet Economy; Western and Soviet Views* (New York: Appleton-Century-Crofts, 1963).

15. R. Spector, Letter, *The New York Times*, Aug. 12, 1986.

16. S. Tefft, "China Is Under Pressure to Clean Up Its Statistics," *The Christian Science Monitor*, June 9, 1994.

17. *The New York Times*, " Rigging the Numbers," June 15, 1998, A22.

18. V. Toy, "Draft Audit Says Board of Education Underrates Crime in Schools," *The New York Times*, Sept. 2, 1995.

19. T. Weiner, "General Details Altered 'Star Wars' Test," *The New York Times*, Aug. 27, 1993.

22

Relevance: Which One is the *Right* Number Trail?

22.1   Preface

Reliable data are not always relevant, that is, well-chosen in connection with the intent of a study being undertaken. If interest centers on the heights of those in a certain community and we are presented with their weights, very carefully obtained, then the data are reliable, but hardly well-chosen in connection with the focus of the study. Subsequent mathematical refinements or conclusions obtained from poorly chosen data might have the aura of *precision*, but are no better than its basic stating points.

Data may be described by various numerical characteristics and the problem of determining which characteristic is "most suitable" for the situation at hand is both challenging and serious.

Another dimension takes us a further step back. A decision-making framework is to be set up for some situation, let us say. The framework requires data. The data obtained may be consistent with the decision-making framework, but if that framework is poorly formulated, the data it leads to cannot be viewed as "well-chosen." This dimension of suitability of data is explored in Case 6.

These considerations, which in my judgment should take center stage in our teaching of mathematics (statistics in particular) are, with perhaps very few exceptions, given no attention in current math education. With one exception, [1], this dimension is ignored in statistics books I am familiar with. It's much less of a drain on our grey cells to turn on the computer and let it do its thing.

The following discussions and examples might be helpful to you if you wish to introduce this dimension to your classes.

22.2   How Relevant are These Numbers?

Case 1   Temperature vs. Wind Chill Factor

Richard Browne was informed by Metro Weather that the temperature was 60°. He put on his jacket and stepped outside, intending to take a walk, but within two minutes he was back inside, shivering. Nobody said anything about that wicked wind, he thought and, turning on the Weather Channel, learned that the wind chill factor was 28°.

The reliability of these numbers is not at issue. The wind chill number is clearly more relevant to the question of what kind of outerwear Richard should use.

Case 2   Which Data "Best" Reflect Airline Reliability?

The long time standard measure of an airline's reliability is its percentage of on-time arrivals, where a flight is deemed on-time if it arrives within 15 minutes of its scheduled arrival time. Such data are widely trumpeted by airlines in their advertising campaigns.

But is this number the "best" measure of an airline's reliability? According to Julius Maldutis, an airline analyst with Salomon Brothers, the answer is no. Maldutis argues that a much better measure of reliability is the percentage of flight-miles completed. Look at the cancellation rate, which is indicative of a more troublesome situation to travelers than that indicated by the artificial on-time number, says Maldutis. [2]

Case 3   Can Andy Afford a $200,000 Porsche?

For many of us $200,000 is a considerable sum, and if Andy's financial state were anything like ours, we would probably be inclined to say no. There's a **big IF** here, that points to the question that is at the heart of the matter: What is Andy's financial state? Since we don't know, we can only proceed by making some **assumptions**.

Scenario 1.
Andy's after tax assets from salary and investments are approximately $50,000 per year. With a $200,000 obligation against $50,000 in assets per year, it's difficult to see how Andy would be able to avoid defaulting down the road.

Scenario 2.
Andy's after tax assets from salary and investments are approximately $2 million per year. With a $200,000 obligation against $2 million in assets per year, we would probably be inclined to tell Andy to go for it.

The lesson to be learned from Andy's situation is that when it comes to a debt to be carried, focusing on the amount of the debt in absolute terms ($200,000 or whatever) is not the appropriate number trail. The appropriate number trail consists of the ratio of debt to ability to pay expressed by a measure of assets. What about countries?, one may ask.

Case 4   The National Debt: Public Nuisance or Menace?

By the beginning of 1993 the gross national debt of the United States, it was generally agreed, was in the neighborhood of $4.2 trillion, a staggering figure which boggles the mind. If you had to transfer this amount of money in $100 bills from one location to another, you would have to deal with a stack of bills 2670 miles long.

The figure sounds ominous, but here is where disagreement begins. One point of view argues that the figure itself and the rate at which it has been increasing portend catastrophic consequences in the offing. When the debt grows faster than the country's ability to carry it, a breakdown with social, political, and economic upheaval is inevitable, and we are coming dangerously close to this state, this view has it.

But is the total size of the debt the figure we should be giving our first priority? Another view argues that in terms of the state of the economy, we are looking at the wrong figure and that, while debt reduction is desirable, it should not be given top priority and carried out in a "mindless" way since this will severely damage the economy. Its proponents focus on the ratio of publicly held debt to Gross Domestic Product (GDP).
In principle, how different is this situation from the one Andy finds himself in?, one may ask.

For discussion see [3]-[5].

Case 5   Is the Recession Over if the Numbers Say So?

In early 1992 President Bush was, as modern parlance would put it, an unhappy camper. Government statistics showed a mild recession and strong economic fundamentals. Yet business and consumer confidence in the economy had been shaken to an extent that seemed way out of proportion to the statistical signs, and many were blaming George Bush for having missed the wake up call.

"The problem," observes Charles McMillion [6], "is that many of those statistics are wildly misleading."

One statistic concerns **unemployment**. During the 1982 recession, the worst since World War II, unemployment reached 10.8 percent. During the 1991-92 recession it reached 7.8 percent. McMillion notes that the 1991-92 unemployment number looks good by comparison because it mixes two factors, jobs and the size of the labor force, and neglects the fact that the labor force has contracted sharply. "A better gauge," he argues, "is the number of actual goods." Three hundred thousand more jobs were lost in the 1991-92 recession than the 1982 recession, June 1981-January 1983. A larger portion of the jobs lost this time involved higher-wage white collar workers, with ramifications throughout the economy. People working or seeking employment has declined by 1.2 million people in the first 19 months of this recession as opposed to 125,000 in the first 19 months of the 1982 recession. These features are not revealed by the unemployment rate.

Manufacturing output is another statistical indicator of economic health. According to this statistic, using constant output values, manufacturing has remained near 22 percent of America's gross national product since World War II. But, McMillion notes: "Even Commerce Department officials who assigned these values admitted—in the Survey of Current Business last year—that 'only a substantial research effort over many years holds any promise of overcoming . . . formidable statistical problems' with these figures."

The rapid race of technological change makes it virtually impossible to measure 'constant' output over time.

U.S. Competitiveness: A comparison of the gross national product per worker of the United States against that of other major industrial competitors shows the United States to be well ahead of such rivals as Germany and Japan. "The tally depends," McMillion observes, "on the value assigned to the dollar . . . . Most comparisons use theoretical—so called 'domestic purchasing power parity' values that vastly overvalue the dollar."

Clearly, the **assumptions** underlying such statistical economic indicators are unemployment, manufacturing output and U.S. competitiveness must be watched. What must also be watched are the limitations of such indicators, and what they omit which is relevant.

Case 6   How Good Is This Decision-Making Framework?

The president of Ecap University charged his Dean of Administrative Affairs, Michael Russell, with the task of setting up a criterion for running or cancelling course sections that would take into account student needs, be sensitive to maintaining academic quality, address the cost dimension, and be simple to use.

Dean Russell started with the **assumption** that each section, with perhaps a few exceptions, should pay its own way. He set up a course section run-cancel criterion, RC-1, based on the difference between the tuition revenue generated for the section based on student enrollment and the salary cost of the instructor for the section. RC-1 says run the section if revenue minus cost equals or exceeds $5000.

$$R - C \geq 5000$$

Otherwise, it is to be canceled, unless a compelling student need for the course could be established.

How well does RC-1 satisfy the requirements for a section run/cancel criterion stated by the president of Ecap University?

Implementation of RC-1 requires that the tuition revenue R and instructor cost C be determined for each section. Consider one such section, Stat 200 A1, let us say. If $R - C = 6000$ we run Stat 200 A1; if $R - C = 4000$, we cancel Stat 200 A1, unless a "compelling" reason to offer it can be presented.

RC-1 gets the job done, but there is a serious question about whether it's the "best" system for handling course run/cancel decision making in a way that satisfies the president's mandate.

1. **Academic Quality.** Department heads are not free to assign the "best" person to teach a section because the "best" person might be too costly in terms of the $R - C \geq 5000$ condition. A senior faculty member assigned to a course will often carry a much higher cost than a junior, or adjunct colleague. If circumstances lead to changes in teaching assignments, courses that would run under one assignment framework might well have to be cancelled under another. With RC-1 the running or cancellation of courses depends more on how the game of academic musical chairs is played out than on the academic needs for running them, which is an unsatisfactory condition. It is in the overall interest of Ecap U. and, in particular, its students, that department heads be able to assign the most

suitable faculty to courses and make changes when circumstances dictate. RC-1 is **not compatible** with this condition.

2. **Tunnel View.** RC-1 does not take into account the total revenue—total cost picture since reorganization of faculty teaching assignments by itself neither changes the total tuition revenue nor the total cost of faculty salaries.

3. **Data Collection.** The seeming simplicity of RC-1 in terms of data needed for its implementation is deceptive. A good deal of data has to be obtained and evaluated for each section in terms of a particular faculty assignment to the section before a run/cancel decision can be make. This is time consuming, but must be carried out within a tight time frame after class registration has taken place, but prior to the beginning of classes.

Case 7   Mean vs. Median

Which number is a more revealing characteristic of data? The answer, of course, depend on the nature of the data.

This point is one that I emphasize to my statistics students. It's so easy to fall into the compute, compute, compute mode that we become in danger of losing sight of this basic question.

The seven full time faculty members in Ecap University's social science department have salaries (in thousands of dollars) 30, 30, 30, 30, 30, 135, 135. When one of the department members complained to the dean about the low salaries in the department the dean pointed out that there is no justification for complaint because the mean salary level is $60,000. What would be an appropriate clarification for the dean's information?

Mean values are strongly influenced by extreme values. When such are present, as is the case here, the mean does not serve as a good indicator of the nature of the data. In such cases the median, which is not influenced by extreme values, is a better indicator of the nature of the data. The faculty's median salary of $30,000 is a more reliable indicator of the social science department's salary state of affairs.

22.3   Food for Thought Questions

1. Abel Fisher, a financial analyst, was hired to analyze the operations of the accounting department of Arley College and make recommendations on how to improve its financial efficiency. The income of the department is the tuition income of the students being serviced minus costs, primarily salary costs. Mr. Fisher collected data on the class size of each instructor and each instructor's rank and salary. He found that a number of the full professors at the top of the salary scale were teaching classes with a small number of students.

    To improve the income of the department he recommended that teachers at the top of the salary scale be assigned classes which can be expected to have a large number of students.

    (a)   Is the data collected appropriate to the problem under study? Explain.

    (b)   Do you agree that implementation of Abel Fisher's recommendation will improve the accounting department's efficiency from an income-cost point of view? Explain.

2. The Baldwin Insurance Company hired marketing analyst Arnold W. Williamson to develop a strategy to make its car insurance policies more attractive. Mr. Williamson suggested that to reward and encourage safe driving the company offer a 5% discount to policy holders who had been with the company for five years and had not been in an accident. An additional 1% discount would be given for each additional year that had been accident free up to a maximum of 15%.

    Do you agree that the number of years of safe driving is the number to focus on as a measure of safe driving?

3. Dean Michael Russell of Ecap University was asked to develop other criteria for running or cancelling course sections which take into account students needs, is sensitive to maintaining academic quality, addresses the cost dimension, and is simple to use. The dean came up with Russell criterion 2, RC-2, which operates as follows: For a given department, English, for example, determine the average salary cost of the faculty in the English department. Run English section A, let us say, if the tuition revenue based on section A's enrollment equals or exceeds the average salary cost of the

English department for section A by $5000. Otherwise, cancel section A unless compelling student or academic needs can be established. The same system would apply to courses in other departments.

(a)   What data would be needed for the implementation of RC-2?

(b)   What are the merits and disadvantages of RC-2?

4. Another criterion developed by Dean Russell, RC-3, operates as follows: For a given department, determine the average class size for all sections being run by the department for the session in question, excluding such one-on-one activities as independent study and thesis supervision. If this average class size is at least 25, then run all sections with an enrollment of at least 10. If the enrollment of a section is less than 10, then go to RC-2 to make a determination for the section. If the average class size is less than 25, then go to RC-2 for all sections.

Address the same questions (a) and (b) for RC-3 as those posed for RC-2 in 3.

(c)   Is RC-3 an improvement over RC-2? Explain.

5. The exam grades of the nine students in an electrical engineering course were 49, 50, 52, 52, 55, 86, 95, 100, 100. Asked how well the class did by one of the students, the instructor replied: "not bad at all; the means is 71."

(a)   Is the mean the best indicator of how well the class did? Explain.

(b)   If your answer to (a) is no, then what do you believe would be a more suitable indicator? Explain.

6. Which provides a better measure of a company's earnings, Generally Accepted Accounting Principles (GAAP) or Standard & Poors Core Earning's Measure (CEM)? N. Brynes, M. Derhovanesian, "Earnings: A Closer Look," *Business Week*, May 27, 2002; 34-37; P. Krugman, "America's Poor Standards," *The New York Times*, May 17, 2002; A25.

Why is the proposed Standard & Poors Core Earnings Measure a big deal? P. Coy, "Why Better Numbers Really Matter *Business Week*, May 27, 2002; 36-37; "Editorial: A Good Idea About Earnings," *Business Week*, May 27, 2002; 114.

7. Which is the better measure of a basketball player's scoring efficiency, shooting percentage or points per shot? How so? T. Keegan, "New Stat Creates Great Divide: Points Per Shot, A Terrific Gauge," *The New York Post*, Feb. 9, 1997; 92.

8. The trade deficit is way up. Should we push the panic button? M. Weidenbaum, "Trade Deficit: Obsessing Over a Misleading Indicator," *The Christian Science Monitor*, June 18, 1998; June 18, 1998; 11 R. Eisner, *The Great Deficit Scares* (New York: The Century Foundation Press, 1997), Ch. 3.

9. "Who would you say is the richer on average: Americans, Frenchmen, Italians, or Germans?" GDP (Gross Domestic Product) vs. GPI (Genuine Progress Indicator); D. Francis, *The Christian Science Monitor*, July 1, 2004; p.17.

10. G.D.P "has become probably our most commonly cited economic indicator. The basic number that we take as a measure of how well we're doing economically from year to year and quarter to quarter. But it is a miserable failure at representing our economic reality." How So Alternatives? Eric Zencey, "GDP. RIP", *The New York Times*, Aug. 10, 2009, p. A15. Is emphasis on growth misguided? P. Goodman, "Emphasis on Growth Is Called Misguided," *The New Yord Times,* Sept. 23, '09; B1, B5; *The Economist*, "Economics Focus: Measuring What Matters," Sept. 19, '09; p.88.

11. "At least one further difficulty inherent in this measure "[G.D.P.], D. Vanderpool, "G.D.P. Flaw; Not All Economic Activity is Productive", (letter in response to Zencey, 10.) *The New York Times*, Aug. 17, '09.

22.4   References of Interest?

1. W. J. Adams, *Statistics: Basic Principles and Applications*, Revised Second Edition (Xlibris: 2009) ch. 1.

2. A. Bryant, "A Different Gauge for Rating Airlines," *The New York Times*, March 7, 1995.

3. J. Davidson, W. Rees-Wogg, *The Great Reckoning*, rev. ed. (New York: Touchstone Books, 1994)

4. R. Eisner, *The Misunderstood Economy* (Boston: Harvard Business School Press, 1994).

5. R. Kuttner, *The End of Laissez Faire: American Economic Policy After the Cold War* (New York: Random House, 1991)

6. C. McMillion, "Facing the Economy's Grim Reality", *The New York Times*, Feb. 23, 1992.

23

Methods for Obtaining the Numbers

23.1 Preface

Far too little attention is given in mathematics education (particularly statistics) to methods for obtaining data. With polls/surveys playing the prominent role they now enjoy in shaping opinions and decision making it is more essential than ever that this dimension be given the attention it deserves, especially in light of Ben Wattenberg's observation, 'one can defend almost any position on any subject by referring to public opinion polls.' [13]

I give it a good deal of attention in teaching statistics and I thought you might find the approach I employ of interest.

23.2 Bull's Eye vs. Off the Target

The polling Goliath that emerged in the period between the two world wars was *The Literary Digest*, a popular weekly magazine of the period. In early 1920 the *Digest* mailed over eleven million ballots to obtain a sense of the public's view on possible presidential candidates. In 1924 it conducted its first presidential poll, sending out some sixteen and a half million ballots to people in all 48 states. For the 1928 and 1932 elections even more ballots were sent out. The mailing list for the ballots came from telephone directories, automobile registration lists, and voter registration lists. In 1932 *The Literary Digest* Poll predicted that Franklin Delano Roosevelt, first time candidate for president, would win the popular vote with a margin of 60% and carry 41 states with 474 electoral votes. Roosevelt received 59% of the popular vote and carried 42 states with 472 electoral votes. This degree of closeness of the predicted values to reality's mark made *The Literary Digest* Poll the poll to be relied on.

1936

The 1936 presidential election was on the horizon with Franklin D. Roosevelt, seeking reelection, facing the Republican nominee Alfred E. Landon and minor party candidates. Riding high, *The Literary Digest* geared up. It mailed some ten million ballots to prospective voters and eagerly awaited the returns.

In response to its poll, the *Digest* received 2,376,523 responses, with 1,293,669 for Landon, 972,897 for Roosevelt (42.9% of the two-party vote) and the remainder for third party candidates. The *Digest* had Landon carrying 32 states with 370 electoral votes and Roosevelt carrying 16 states with 161 electoral votes. The actual vote gave Roosevelt a landslide victory with 62.5% of the two-party popular vote. He carried 46 instead of 16 states and received 523 instead of 161 electoral votes.

So close and so far. The *Literary Digest Poll* went from being within 1% of reality's mark in 1932 to being off by 19% of reality's mark in 1936 in predicting the two-party popular vote for the winner of the presidency.

What lessons can we learn from this debacle?

23.3   Lessons

Sample Size: The accuracy of a poll is not determined by enormous sample sizes, impressive as they might appear to be. The major point is that the sample must be "properly" chosen. Pre-election poll samples these days tend to be between one and two thousand in size.

Target Population vs: the Sampled Population: It is most important that the population actually being sampled be the one about which we seek to draw inferences. The failure of *The Literary Digest's* 1936 poll was in part due to the sampled population being significantly different from the target population of prospective voters. It was generally appreciated that samples drawn largely from telephone directories and automobile registrations favored wealthy Americans who tended to vote Republican, but the extent to which this might be the case was not generally appreciated. After all, the *Digest's* spectacularly successful 1932 poll drew from the same sort of lists.

What went wrong? In 1932 voters of all economic *strata* tended to vote against President Hoover and the Republicans, holding them responsible for the Great Depression and its aftermath. Economic class differences were obscured, and *The Digest* got lucky in its poll. In 1936 the upper economic *strata*, disturbed by the direction of President Roosevelt's *New Deal,* were much more willing to return to the Republicans, as reflected by the *Digest's* 1936 poll. Spectacular success without an understanding of its origins bred spectacular arrogance followed by spectacular failure.

Non-response Error: Of ten million odd ballots sent out by *The Digest*, 2,376,523 returns were received for a 76% non-response rate. What were the non-respondents thinking? It is hazardous to predict one way or another. Those who did respond wanted to make sure that their opinion was counted. Those who did not respond did not care about the poll. In some locations, such as Allentown, Pennsylvania, the sampling list was not drawn from telephone and automobile registration lists, but from voter registration lists with no inherent Republican bias. The data show a clear response bias in that the proportion who favored Landon in the poll was much higher than the proportion who favored him in the election. Those voters who came from the lower economic *strata* tended not to respond to *The Digest* poll when invited to do so and they strongly favored Roosevelt.

23.4   The Polling Goliath and the Three Davids

If the title *Goliath of Polling* existed in 1932 the Literary Digest Polling Organization would have been so crowned on the basis of the incredible success of its 1932 presidential poll.

Against this background it seemed foolhardy in the extreme for George H. Gallup to take on Goliath. Gallup had founded the *American Institute of Public Opinion* in 1935 and initiated a weekly column called *America Speaks*. To attract newspaper subscriptions he offered a money back guarantee that his prediction of the presidential winner would be more accurate than that of *The Literary Digest* in the 1936 election.

Gallup appreciated that a survey sample should be representative of the voting population at large and he employed quota sampling to obtain an appropriate sample mix. He also appreciated the problem of nonresponse bias and sought to minimize it. Gallup predicted a Roosevelt victory with about 54% of the popular vote.

Independently of Gallup, two other researchers, Elmo Roper and Archibald Crossley, using similar methods, had also predicted a Roosevelt landslide. The three Davids, particularly Gallup, and their methods stood tall over the fallen Goliath humbled by the reality of the 1936 election.

Mail Polls: A mail poll, employed by *The Digest*, carries with it a seed with the potential to render it useless; the seed is nonresponse bias. Gallup was much closer to the mark than the *Digest*, but his predictions were wrong on six states and he had predicted a Roosevelt victory with about 54% of the populate vote, seven percentage points off the mark. One lesson for Gallup was that mail balloting was too unreliable for pre-election polling.

Other lessons were to come, but the laboratory of reality required more time to make clear that additional fine tuning was required. Gallup, Crossley, and Roper were pioneers in the quota method of sampling which they carried out effectively enough to obtain far more accurate predictions than the *Digest* achieved by its mail poll, but still left them significantly short of the mark. It might not matter when the outcome is strongly one-sided, but in a close election it could make a big difference. The Dewey versus Truman election of 1948 provided a decisive test case.

23.5   The 1948 presidential Election: Further Lessons

Thomas E. Dewey vs. Harry S. Truman

Vice President Harry S. Truman became President on the death of President Franklin D. Roosevelt on April 12, 1945. It was a difficult post-war period and when Truman was nominated by the Democratic Party as its presidential candidate in 1948 he faced an uphill fight against the Republican challenger Thomas E. Dewey, Governor of New York. The polls showed Dewey comfortably ahead and he adopted a strategy of caution and platitudes ("Our future lies before us."), seeking to avoid making commitments and enemies. The political establishment, with the exception of Truman, felt the Dewey had it in the bag. Truman vigorously counterattacked, delivering 300 odd "give'em hell" speeches on a whistle-stop tour of the country.

Truman went to bed early election day evening and woke up to find himself President for another term. The pollsters and political pundits had missed the boat. The actual popular vote and the predictions of Crossley, Gallup, and Roper are shown in Table 23.1.

Table 23.1

	Dewey (%)	Truman (%)	Other Candidates (%)
Actual vote	45.1	49.5	5.4
Crossley	49.9	44.8	5.3
Gallup	49.5	44.5	5.5
Roper	52.2	37.1	10.7

It was a close election, no doubt. In electoral vote terms, Dewey would have won had he carried California, Illinois, and Ohio, each of which he lost by less than 1%. The size of Gallup's and Crossley's error in predicting Dewey's victory in 1948 (5.4 and 4.7 percentage points, respectively) was smaller than the error in predicting Roosevelt's victory in 1936 (7 percentage points for each). But in terms of the bottom line wrong is wrong, and the exaggerated faith that had come to be placed in "modern scientific polling," as it had come to be called, was shattered. What happened?

Further Lessons

Stability of Voter Opinion

By early September the polls showed Dewey ahead of Truman by at least ten percentage points. The pollsters believed that public opinion was pretty well set by this time and would change little in the time remaining before the election. At this point they closed down their polling operations. Reality, of course, proved them wrong.

The Undecided Vote

How is the undecided vote to be handled? Gallup split the undecided vote in the same proportion as those who had expressed a preference for Dewey and for Truman, which strongly favored Dewey. Subsequent analysis showed that 14 percent of the voters made up their minds in the last two weeks of the campaign and that 74 percent of these went for Truman. It is clearly hazardous to decide for the undecided.

Quota vs. Probability Sampling

Gallup and other pollsters were criticized in 1944 for continuing to use the quota method[1] in their polling. Their failure in 1948 brought matters to a head. Stratified random sampling[2] was the technique favored by many critics, but it was more complicated and much more expensive to implement. Gallup had serious doubts that the increased accuracy achieved would be worth the expense, but after the 1948 election he and other pollsters switched to stratified random sampling for pre-election polls.

---

[1] In **quota sampling** the interviewer is instructed to interview an assigned number, or quota, of individuals in groups defined by specified characteristics (race, religion, occupation, for example). The choice of who within the quota is to be interviewed is made by the interviewer.

[2] **Stratified random sampling** involves dividing the population to be sampled into subpopulations called *strata* (such as states) and taking a random sample from each *stratum*.

Background is interesting and important, but how do things stand now? This is addressed in the sections that follow.

23.6   Polling and Its Perils

Aftermath of the 1948 Election

Faith in polling as a reliable instrument for measuring public opinion was shaken by the scope of its failure in the 1948 election. But, as we see from the scope of its use in current affairs, that faith has been restored with some reservations. A better awareness of difficulties on the part of professional pollsters who are concerned with accuracy and reliability has led to important refinements in the art and science of survey design, interviewing, and poll taking.

The Affect of the Instrument

The problem of obtaining an accurate poll reading is analogous to that of obtaining an accurate blood pressure reading. The presence of the instrument itself affects that which is being measured. In determining a person's blood pressure the white-coat syndrome refers to the tendency of blood pressure to rise when the person is being examined by a medical professional. There are many white coats in survey or poll taking and we look at several of them here from the point of view that to be aware of problems is the first step towards dealing with them.

The Poll Itself

Being polled itself may put people on guard if they feel that they might loose benefits or be penalized in some way if they give the "wrong" answers. In the 1930's for example, many people on welfare were afraid they would be thrown off the welfare rolls if they gave an undesirable response. The election of Violeta Chamorro as President of Nicaragua in early 1990 was contrary to poll predictions which had Daniel Ortega with a substantial lead. One reason that the polls were so inaccurate was that the intimidation factor was not accurately taken into account. Nicaragua had been under authoritarian rule and in a state of civil war for a number of years, and many voters were not about to freely express their political preferences to pollsters. In such situations it is especially important to go all out to win the trust of respondents.

Interviewer Induced Bias

Conducting a "successful" interview, whether in-person or by telephone, is not a simple matter. People respond to interviewers as well as questions and the interviewer must strike a balance in being personable, respectful and considerate of the person being interviewed, and professional. The type of person who sets your teeth on edge by the way he says Good Morning is not likely to be a successful interviewer.

Questions

Scope of Questioning. In any situation involving questioning it is not just a matter of the questions asked but also those not asked that add their coloring to the picture we obtain. If, for example, the presidential polls of 2000 had focused exclusively on the two major party candidates, George W. Bush and Albert Gore, Jr., they would not have seen the impact that Green Party candidate Ralph Nader was having on the election's outcome.

Nuances in Wording. The wording of poll questions is, with the best of intentions, a delicate matter. In an experiment conducted by Elmo Roper as far back as 1940, of those asked if the U.S. should forbid public speeches against democracy, 46% replied "no." Of those asked if the U.S. should allow public speeches against democracy, 25% said "yes." Support for free speech was much greater when the term "forbid" was used rather than "allow."

In the fall of 1997 the following question appeared on a referendum ballot in Houston. "Shall the charter of the City of Houston be amended to end the use of affirmative action for women and minorities?" The initiative was defeated by a 55-to-45 percent margin. But is its wording equivalent to that taken from the 1964 Civil Rights Act? "The City of Houston shall not discriminate against, or grant preferential treatment, to any individual or group on the basis of race, sex, ethnicity, or national origin." Judge Sharolyn Wood argued that it was unfair to change the original wording, threw out the results of the Houston referendum, and ordered a new vote.

Meaning of Words. "Do you think President Nixon should be impeached and compelled to leave the Presidency, or not?" In a Gallup poll held in July 1974, 24% of the respondents said "yes." When the question was posed as, do you "think there is enough evidence of possible wrong doing in the case of President Nixon to bring him to trial before the Senate, or not?", 51% said "yes." This was in response to a Gallup poll held at the same time. A case of fickle respondents? No; the second wording makes clear that Nixon would be brought to trial, which was not as clear to respondents from the first wording because the meaning of "impeached" was in doubt.

Questions involving terms that might be understood one way in everyday usage, but have a specific technical meaning, must be handled carefully. Robbery, involving confrontation between victim and offender, for example, is technically different from burglary, which does not involve personal confrontation. Such a distinction must be made clear in any question involving such terms.

Poorly Formulated Questions can, not unexpectedly, bring an ambiguous response. This question was asked in a 1992 poll conducted by the Roper Organization for the American Jewish Committee: "As you know, the term 'Holocaust' usually refers to the killing of millions of Jews in Nazi death camps during World War II. Does it seem possible or does it seem impossible to you that the Nazi extermination of the Jews never happened?" It's not unreasonable to expect the double negative to "seem impossible" . . . "never happened" to cause confusion, and this is what the responses received reflect.

A study released by the Education Department in September of 1993 concluded that half of the adults in the United States cannot read or handle arithmetic. This is certainly an alarming figure, but is it accurate? That, to a large extent, depends on the success of the test makers in developing questions free of cultural bias, verbal ambiguity, and distracting irrelevancies. The following arithmetic question was posed:

"The price of one ticket and bus for 'Sleuth' costs how much less than the price of one ticket and bus for 'On the Town'? A charter bus will leave from the bus stop (near the Conference Center) at 4 p.m., giving you plenty of time for dinner in New York. Return trip will start from West 45th Street directly following the plays. Both theaters are on West 45th Street. Allow 1 and 1/2 hours for the return trip. Time: 4 p.m., Saturday, November 20. Price: On the Town, Ticket and bus: $11.00; Sleuth: Ticket and bus: $8.50. Limit: Two tickets per person."

The question itself raises the question of to what extent it is intended to test arithmetic and to what extent it is intended to test one's ability to successfully negotiate a verbal maze, particularly when many of those participating in the test are foreign born adults whose first language is not English. With tests of this sort, the underlying question of how well the questions achieve their intended objective must be given careful consideration before meaningful conclusions can be drawn about the population being studied.

Loaded Questions. And then there are loaded questions worded to elicit a desired response for political purposes. Ross Perot's mail poll on National Referendum—Government Reform which appeared in the March 20-26, 1993 issue of TV Guide contains questions of this type. Question 13, for example, reads: "Should laws be passed to eliminate all possibilities of special interests giving huge sums of money to candidates?" The term "special interests" as it is used carries with it the ominous suggestion of special interests taking over the country. When stated in this form in a Time/CNN poll, it received an approval rating of 80 percent. The Time/CNN poll also put the question as follows: "Should laws be passed to prohibit interest groups from contributing to campaigns, or do groups have a right to contribute to the candidates they support?" Forty percent of the Time/CNN respondents stated they would prohibit interest groups from contributing, while 55 percent stated that they had a right to contribute.

Underloaded Questions. Can an unloaded question be loaded by virtue of being unloaded? There are those who would say yes in connection with a propositions to be put before California voters in November 1996. The proposition reads: "The state will not use race, sex, color, ethnicity or national origin as a criterion for either discriminating against, or granting preferential treatment to, any individual or group in the operation of the state's system of public employment, education, or public contracting." The proposition does not mention affirmative action, a term which evokes a strong emotional response, but if passed it would end affirmative action programs in California.

When put as stated to a representative sample of 800 California voters in a Harris poll conducted in May 1995, they favored it by 78 to 16 percent (6% undecided). When it was pointed out that the proposition would outlaw all affirmative action programs for women and minority groups, the pros of the sample went from 78 to 31 percent and the cons from 16 to 55 percent (14% undecided).

Direct Questions may provide a misleading response. Do you intend to vote in the forthcoming election? Many people would answer yes rather than run the risk of being thought an irresponsible citizen.

Personal Questions concerned with drug use, one's sexual behavior or preferences, for example, may provoke a response considered socially acceptable rather than candid.

It was recently found that computers elicit more honest responses to delicate personal questions that human beings [5] and [12]. The more socially stigmatized the behavior, the greater the difference in response, it was found. This finding threatens to call into question data that had previously been obtained on a number of sensitive subjects.

Question Order. Subtle differences in question ordering can have a significant impact on the responses obtained. The January 1984 pre-presidential *New York Times*/CBS poll found that voters preferred incumbent President Ronald Reagan to Democratic challenger Walter Mondale by 16 percentage points; the question was posted at the beginning of the interview, which favored the better known Reagan. Gallup and *Washington Post*/ABC News polls taken around the same time posed the question near the end of the interview after the questions about Reagan's policies had been asked; this helped Mondale because he had less of a record to defend. These polls had Reagan and Mondale about even.

Response Options

The two-category response option of the form Agree or Disagree, for example, is much more restrictive than the four-category response option Agree, Disagree, Not sure, Not enough information. The number of response options available may profoundly influence the response given, including the possibility of nonresponse.

It is interesting to compare voter reactions in two July 1992 opinion polls on President George Bush and Democratic Party candidate Bill Clinton. The responses for the two polls are shown in Tables 23.2 and 23.3.

Table 23.2

	Bush	Clinton
Favorable	40%	63%
Unfavorable	53%	25%
Don't know	7%	12%

Table 23.3

	Bush	Clinton
Favorable	27%	36%
Unfavorable	49%	24%
Undecided	22%	31%
Haven't heard enough	2%	9%
No answer	1%	1%

KISS vs. Complex

There is merit to the KISS (Keep it simple, stupid) philosophy in that keeping it simple avoids potential bias and misunderstanding which may arise from complex questions. It is one matter to ask interviewee Arnold Mount who he favors in the forthcoming election and another to present Mount with a list of a hundred questions on domestic and foreign policy. Of course, the less one asks, the less one will learn, and there is an important balance between simple and simplistic that must be kept in mind.

Long and/or Loaded. Greg Schneiders and Jo Ellen Livingston note that Microsoft claims that the results of a survey it commissioned indicate overwhelming support for incorporating its Internet browser into the Windows platform. "The 350-word question," they point out, "mostly argued in favor of incorporating the browser; nothing in the question argued against it. What's remarkable is that 15% of respondents could actually bring themselves to disagree." [8]

The Unavoidable Gap

"There's many a slip between the cup and the lip," the old saying has it, and this is especially the case with the gap between the last poll taken and the election itself. A poll may be likened to a picture taken by a camera at a point in time. Looking at a sequence of polls is analogous to looking at a sequence of still-photos, a videotape, if you will. While this might be strongly indicative of what will happen, it is not the same as what does happen.

23.7   Qualitative vs. Statistical Studies

Sexuality By the Numbers, or Not?

In 1987 Shere Hite published *Women and Love: A Cultural Revolution*, her third book on human sexuality. In her first two books, *The Hite Report* (1976) and *The Hite Report on Male Sexuality* (1981), Hite restricted herself to telling what she had learned from women and men who replied to the extensive questionnaires concerning their sexual problems and attitudes she had circulated. For her first book she circulated approximately 100,000 questionnaires, from which she received 3019 responses for a response rate of about 3%. Four different versions of the questionnaire were sent to women's organizations that were asked to circulate them. A similar methodology was employed for her second book. Approximately 119,000 questionnaires were distributed with a response rate just under 6% being obtained.

Shades of the *Literary Digest's* debacle come to mind, but Hite was not claiming that her sample was representative of women and men in general. Hers was a qualitative rather than statistical study. In statistical studies, the same questions must be asked of all prospective respondents with the same response options being available to them all. Uniformity of the underlying conditions and the choosing of a "representative" sample so that the results obtained could be projected onto the population at large are essential for a statistical study. Qualitative studies, on the other hand, focus on the special qualities of each individual potential respondent. Capturing the diversity inherent in individuals takes priority over ensuring uniform underlying conditions. It is not a matter of one kind of study being superior to the other, but rather of which methodology is appropriate to the study being undertaken.

Although her first two books raised much controversy, Hite was on safe methodological ground.

In her third book, Hite attempted to cross the bridge from the qualitative results she had obtained to statistical generalizations about sexual attitudes of women in America. The bridge collapsed. Her methodology was almost universally criticized. ABC News in conjunction with *The Washington Post* conducted a telephone poll for October 15-19, 1987 to see if they could duplicate her results. They could not; their results were sharply at variance with those projected by Hite [6]. To take two examples, she found 84% of women as not being satisfied emotionally with their relationships; ABC/ WP found 7% of married women and single women in a relationship as not being emotionally satisfied. Hite found 78% of women feeling they are only occasionally treated as equals most of the time. The ABC/WP figure was 9%. There were differences in the way questions were posed, but they were not startling. As to who is closer to the mark, ABC/WP clearly takes the Trustworthy Prize because of its sound statistical methodology.

23.8   How Trustworthy Are These Poll Results?

Caution:

Keep the following in mind:

1.   Context and Basic Information

Focusing on poll results without an appropriate context is misleading. News accounts of poll results should give information about the date the poll was taken, sample size, survey design, percentage of respondents among those contacted, response options, random sampling error and what it means. Out-of-context poll results are best viewed with questioning skepticism.

2.   Questions.

The complete wording of questions should be provided. We may then ask ourselves: Are the questions clearly posed? In addition to the questions posed, are there questions that should have been posed to avoid overall bias in the questioning? Are there leading questions whose coloring would favor a certain kind of response? Are there personal questions that a respondent might be reluctant to answer truthfully? At first thought it might not seem like a big deal to ask a person his preference in an upcoming election, but where an atmosphere of repression or intimidation exists it is indeed a most sensitive question.

---

3.   The Response Rate.

What was the response rate? A major lesson of *The Literary Digest* 1936 presidential is that a low response rate may render the results obtained untrustworthy as a basis for predicting the attitudes of the target population. This issue is more complex than it might seem to be at first sight. For further discussion, see G. Langer, "About Response Rates: Some Unresolved Questions," *Public Perspective*, May/June 2003; 16-18.

---

4.   Self-Selected Respondents.

A popular, but seriously flawed, polling technique is the mail-survey, magazine, online, or telephone poll in which the public is invited to respond to a written questionnaire, or call one number to register approval of a candidate or position and another number for disapproval. Polls of this sort lend themselves to gross manipulation by individuals and pressure groups and give us no handle on the opinions of non-respondents. Self-selected respondents represent only themselves.

Any claim or suggestion that results obtained from polls of this sort express the views of a wider population is total without merit.

---

5.   Online Polls.

The first stage of online polling does not differ from mail-in or telephone-response polls. Pseudo-poll would be a more appropriate label for such techniques rather than poll, which has come to suggest a rigorous, scientific framework not possessed by the first stage of online polls and their ilk.

The advantage of online pseudo-polls over their kind is that the internet dimension permits the development of this pseudo-poll into a "legitimate" poll, sometimes termed an interactive online poll. This development is in progress.

When presented with the results of an online poll there is no way of knowing whether they are from a pseudo-poll or interactive online poll unless the methodology is identified. This is often not done.

Beware.

---

6. Who Commissioned The Poll/Survey and Who's Doing It?

Many honest polls/surveys are commissioned by interested parties, but there are fair-minded interested parties and not-so-fair-minded interested parties who are more concerned with manipulating public opinion than obtaining an objective assessment of it. It is useful to know who wants to know and who's doing the survey to help provide some perspective for the results obtained.

---

7. Qualitative vs. Statistical Studies.

Each type, in its own right, may yield valuable information and insights. Equating the results of the two types of study results in nonsense and muddled thinking.

23.9   Is There More to It Than What the Numbers May Suggest?

That this is the case should come as no surprise. Numerical data, even what may seem to be of the simplest kind, often require qualifications to add to their reliability in the sense of completeness. Metro Weather reports the temperature reading at Smithtown as 60°, for instance. The Reading is 60° at the location it was taken, but the value reflects the physical conditions at the particular time it was obtained, conditions that are not reported. When it comes to more complex data such as that obtained from a poll seeking to determine peoples' attitudes the difficulty in realistically capturing their views becomes immeasurable more difficult. Godfrey Sperling's experience provides us with a valuable lesson.

"It's much safer to stay here in Washington and rely on the polls for your readings for what the people out there actually are thinking. Do that and you will remain convinced that Americans are bored to tears with Clinton scandals and almost completely absorbed in enjoying their own lives," observed veteran political columnist Godfrey Sperling in an article published in June of 1998 [10]. But Sperling ventured out, once to the West Coast and then to the Midwest.

> During that time I listened in on or was part of a number of conversations. And I also asked questions of many people . . . . my reading, unscientific of course, but at least gained from first-hand reporting, is that the polls are wrong . . . . I've now found a public that may be telling pollsters this while really keeping a close eye on and intense interest in the Washington Scene . . . . Yes, whether I'm entitled to or not I've become convinced that there's no new lasting public tolerance emerging—one that will put up with anything this president is accused of doing, even if there is strong indication that he has done it. There is latent outrage out there, among those who keep saying, over and over again, how bored they are with what's going on in Washington.

23.10   Food for Thought Questions

1. The City Council of Bell City wants to obtain a sense of the public's view of the effectiveness of the City's Medical Emergency Response Team (MERT). A random sample of 200 residents was chosen from the City's home owners listing and sent a questionnaire. One hundred twenty five responses were received; 95 gave MERT a favorable rating and 30 gave MERT an unfavorable rating. The City Council concluded that MERT has a 76% favorable rating by the public and, with much satisfaction, announced this result to the news media. The intent of random sampling is to draw the sample in an unbiased manner which does not favor certain samples being drawn over others.

   (a)   Did the polling organization that conducted the poll choose the appropriate target audience? Explain.

(b)   If your answer to (a) is no, what target audience would be appropriate? Explain

(c)   Do you agree with the City Council's assessment that MERT has a 76% favorable rating by the public? Explain.

2. Based on the results of the *Christian Science Monitor*/TIPP poll conducted Oct. 7-13, 2002, it was concluded that "seventy-five percent of Americans say it's important that the U.S. take military action against Iraq by April." The question posed and the results were the following:

How important do you think it is for the U.S. to take military action within the next six months in order to remove Saddam Hussein from power in Iraq?

Very important:	46%
Somewhat important:	29%
Not very important:	13%
Not at all important:	9%
Not sure/Refused:	3%

(a)   Do you Agree or Disagree with the view that this being the only question posed concerning Iraq and the nature of its wording rig the poll to favor going to war with Iraq? Explain.

(b)   Are there questions and response options that you would recommend be added to the survey to obtain a more accurate assessment of the public's views? If so, state them.

(c)   Would you accept the afore figures as accurate without additional information being give? Answer Yes or No and explain the basis for your view.

(d)   If you answered No, what additional information would you want provided before you would feel comfortable reaching a decision about the cited figures? Explain.

3. The Center for Critical Thinking on Domestic and World Affairs is preparing a questionnaire to obtain a sample of public opinion on

domestic and world affairs. The following is a sample of some of the questions being posed.

1.   Would you vote for a presidential candidate who was willing to take more out of your pocket by raising taxes?

2.   Do you want the nation's defense capability reduced by budget cuts in an age of rampant terrorism?

3.   Do you believe that very high priority should be given to reducing the crushing budget deficit that has been imposed on our country?

4.   Do you believe that we should continue to squander money on foreign aid while there are so many urgent domestic needs that require attention?

The Center is engaged in pre-testing its questionnaire. If you were approached, what opinion would you give them about the four questions noted and the response options?

4. Return to the setting of question 3 involving the Center for Critical Thinking. Two response options, Yes and No, are envisioned.

   (a)   Is this sufficient for obtaining an accurate assessment of respondents' views? Explain.

   (b)   If your answer to (a) is no, what response options would you suggest be added? Explain.

5. *The Huxley College Press,* which wants to obtain a sense of faculty opinion about Huxley College's budget allocations for the forthcoming year, sent a reporter to the School of Business to interview a randomly chosen sample of faculty in the accounting, finance, marketing, management, and business law departments about this matter. Will doing this give the *Press* what it wants? Explain.

6. Mayor Keith Joos of Masters-on-the-Mississippi is planning to run for re-election. On a recent talk show he invited the public to call a 900 number and express a favorable or unfavorable rating of his administration. Five hundred calls were received; 350 callers gave his honor an unfavorable rating and 150 gave him a favorable rating. With a 70% unfavorable rating, he strongly considered not running for re-election. Is his pessimism over the results of this call-in warranted? Explain.

7. What factors contributed to the failure of the Crossley, Gallup, and Roper polls to correctly predict the outcome of the 1948 presidential election? Explain.

8. "Rate Your Professor" is an online means for students, to state and obtain views on professors. Although intended for students, anyone who chooses to may participate. Discuss the pros and cons of this means for obtaining information about professors.

9. In the article "Powell in The Middle," by Michael Hirsh and Roy Gutman, (*Newsweek*, Oct. 1, 2001, 26-29), the following assertion is made in boldface without further information being given:

    "71% favor striking terrorists bases even if civilians die, but 59% say we should take time to plan a response that will work."

    (a) Would you accept the figures as accurate without additional information being given? Answer Yes or No.

    (b) If you answered Yes, explain the basis for your view.

    (c) If you answered No, what additional information would you want provided before you would feel comfortable reaching a decision about the cited figures? Explain.

10. "I don't understand what went wrong," lamented an editor of *The Literary Digest*. "We were so on-target with our 1932 poll and so off-target with our 1936 poll, for which we used the same methods." What would you tell him?

11. Does use or non-use of the big A words, Affirmative Action, in a question make a "significant" difference in how respondents reply? L. Harris, "Affirmative Action and the Voter," *The New York Times*, July 31, 1995; A13. J. Leo, "Hold the 'Wrong' Story," *U.S. News & World Report*, Aug. 10, 1998; 12.

12. To obtain a "sense" of what an MBA is "really" worth *Business Week* surveyed the class of 1992 from the "Top 30" MBA programs. BW reached about 4800 alumni out of a total graduate pool of approximately 5700; one thousand four hundred ninety six (1496) responded. For discussion of BW's conclusions and a description of how the survey was carried out see J. Merritt, K. Hazelwood, "What's an MBA Really Worth?," "How We Conducted the Survey," *Business Week*, Sept. 22, 2003; 90-98.

    Do you have concerns about the reliability of the picture painted by the responses received? Explain.

13. In failing to present non-military options to the public, are the pollsters reflecting public opinion or shaping it? F. Solop, K. Hageb, "War or War? 9/11 Surveys Restricted the Options," *Public Perspective*, July/August, 2002; 36-37. What did Thomas Friedman find when he talked to a number of people about Iraq as opposed to what the polls suggested? T. Friedman, "Iraq Upside Down," *The New York Times*, Sept. 18, 2002; A31.

14. The image of sex in America communicated by much of the popular media is one dominated by extramarital affairs and rampant casual sex.

    J.K. Wilson, "U.S. Sex Survey Had Flawed Methodology," *The New York Times*, Nov. 1, 1994; A26.

    Is there a question about the reliability of the figures? See:

    R.C. Lewontin, "Sex, Lies, and Social Science," *The New York Review of Books*, April 20, 1995; 24-29.

    R.C. Lewontin *et al.*, "Sex, Lies, and Social Science: An Exchange," *The New York Review of Books*, May 25, 1995; 43-44.

R.C. Lewontin *et al.*, "Sex, Lies, and Social Science: Letters," *The New York Review of Books*, June 8, 1995; 68-69.

R.C. Lewontin *et al.*, "Sex, Lies, and Social Science: Another Exchange," *The New York Review of Books*, Aug. 10, 1995; 55-56.

M.G. Lord, "What that Survey Didn't Say," *The New York Times*, Oct. 25, 1994; A22.

15. President Clinton's pre-election poll numbers were inflated? What was the problem? R. Morin, "A Matter of Incumbency," *The Washington Post National Weekly Edition*, Dec. 9-15, 1996.

16. What went wrong? R. Morin, "A Bad Night at the Exits," *The Washington Post National Weekly Edition*, Nov. 11-17, 1997; 42.

17. Does it matter who's asking the questions? How so? R. Worcester, K. Kaur-Ballagan, "Who's Asking?," *Public Perspective*, May/June 2002; 42-43.

18. Polls count, but watch out; for what? J. Purnick, "Polls Count, But Could Use a Closer Look," *The New York Times*, Oct. 2, 2000; B1.

19. "The consensus held that the ruling Conservatives would lose their absolute majority in the House of Commons." In fact, the Conservative Party won decisively. What happened? E. Ladd, "Kinnock Defeats Major—Oops," *The Christian Science Monitor*, April 17, 1992. "Why the Polls Got It Wrong Last Time," *The Economist*, March 22, 1997; 68. "The Waterloo of the Polls," D. Butler, D. Kavanagh, *The British General Election of 1992* (New York: St. Martin's Press, 1992), Ch. 7.

20. How does the choice of wording affect how Americans view Iraqi involvement in Sept. 11? T. Zeller, "How Americans Link Iraq and Sept. 11," *The New York Times*, March 2, 2003; wk-3.

21. Upper Slabodia's political establishment wants to launch an attack on Lower Slabodia under the pretext that Lower Slabodia is producing

weapons of mass destruction and is a threat to world peace. One aspect of a public relations campaign to generate public support for the attack is through polling.

Inspired by Ben Wattenberg's remark, "one can defend almost any position on almost any subject by referring to public opinion polls," [13], the political establishment of Upper Slabodia has hired you as an advisor on how to construct the kind of poll that would show that the public favored launching an attack on Lower Slabodia.

(a)   What advice would you give them on questions to be posed, wording, questions to be avoided, response options, timing, etc. Be specific and explain the basis for your advice.

(b)   What advice would you give them on information about the poll's details that should be released to the public and details that should be kept confidential? Explain the basis for your advice.

(c)   As an observer of Upper Slabodia's political scene, you are suspicious of any poll organized and run by the country's media. What questions would you ask about the nature and conduct of the poll?

22. "Election polling had a terrible year in 1996 . . . . viewed against polls' performance recently, the 1948 record seems closer to triumph than disaster," noted Everett Carll Ladd. What is the basis for his assessment? E.C. Ladd, "The Election Polls: An American Waterloo," *The Chronicle of Higher Education*, Nov. 22, 1996; A52.

23. Richard Morin takes issue with Everett Ladd's assessment. On what basis? "The Votes Are Counted, But the Fight Has Just Begun," *The Washington Post National Weekly Edition*, Nov. 25-Dec. 1, 1996.

"Pollster Wars: Round 2," *The Washington Post National Weekly Edition*, Dec. 2-8, 1996; 96.

"Right on the Money," *The Washington Post National Weekly Edition*, March 3-9, 1997; 35.

24. Why, according to Michael Barone, are opinion polls worth less? "Why Opinion Polls Are Worth Less," *U.S. News & World Report*, Dec. 9, 1996; 52.

25. A warning about scurrilous survey techniques? How so? R. Morin, "When Push Comes to Shove," *The Washington Post National Weekly Edition*, July 1-7, 1996; 35.

26. What, according to Harry O'Neill, are the troublesome aspects of polling that pollsters too often look away from?

R. Morin, "Warts and All," *The Washington Post National Weekly Edition*, Oct. 13-19, 1997; 34.

27. One in five Americans doubted the Holocaust happened; can this view be right?

J. Kifner, "Pollster Finds Error on Holocaust Doubts," *The New York Times*, May 20, 1994.

E.C. Ladd, "Setting the Record Straight On a Holocaust-Denial Poll," *The Christian Science Monitor*, July 1, 1994; 18.

M. Kagay, "Poll on Doubt Of Holocaust Is Corrected," *The New York Times*, July 8, 1994; A10.

28. Flawed data from a major survey on food use? What was the problem? What were the consequences? M. Burros, "Major U.S. Survey on Food Use and Pesticides is Drawing Fire," *The New York Times*, Sept. 11, 1991; C1.

29. "Subtle Differences in question order can have a major impact on the results." How so? R. Morin, "What's the Question—and Where?" *The Washington Post National Weekly Edition*, July 28, 1997; 37.

30. How have the Republicans modified their language to appeal to suburban voters? J. Lee, "A Call For Softer, Greener Language: G.O.P. Advisor Offers Linguistic Tactics for Environmental Edge," *The New York Times*,

March 2, 2003; 24N; "Editorial: Environmental Word Games," *The New York Times*, March 15, 2003; A16.

31. Can "seemingly small differences in the wording of survey questions" produce significantly different results? R. Morin, "It Depends What Your Definition of 'Do' Is,' *The Washington Post National Weekly Edition*, Dec. 21-28, 1998; 38.

32. What do some survey's of the Clinton crisis illustrate? R. Morin, "It's All in the Wording," *The Washington Post National Weekly Edition*, Jan. 18, 1999; 21.

33. Is there a difference between what the polls say and what they mean? D. Yankelovich, "What the Polls Say—and What They Mean," *The New York Times*, Sept. 17, 1994; 23.

34. H. Taylor *et al*, How so? "Touchdown! Online polling scores big in November 2000," *Public Perspective*, March/April 2001; 38.

23.11   References of Interest?

1. W. J. Adams, *Statistics: Basic Principles and Applications*, Revised Second Edition (Philadelphia: Xlibris, 2009) The Second Edition was authored by W.J. Adams, I. Kabus, M.P. Preiss.

2. M. Barone, "Are the Polls Accurate?", *The Wall Street Journal*, Oct. 22, 2008, p. A17.

3. G. Gallup, "Opinion Polling in a Democracy," *Statistics: A Guide to the Unknowen*, ed. by J.M. Tanur, *et al*. (Holden Day, 1972), 146-152.

4. G. Gallup, S.F. Rae, *The Pulse of Democracy* (New York: Greenwood Press, 1968)

5. A. Harmon, "Underreporting Found on Male Teen-Ager Sex," *The New York Times*, May 8, 1998.

6. "Hite/ABC Comparison Analysis," news release by ABC for 6:30pm, EST, Monday, Oct. 26, 1987.

7. D. Moore, *The Super Pollsters: How They Measure and Manipulate Public Opinion in America* (New York: Four Walls Eight Windows, 1992).

8. L. Rogers, *The Pollsters; Public Opinion, Politics, and Democratic Leadership* (Alfred A. Knopf, 1949).

9. G. Schneiders, J. Livingston, "Can You Trust the Polls? Well, Sometimes," *The Wall Street Journal*, Feb. 8, 1999.

10. G. Sperling, "Me Against the Pollsters," *The Christian Science Monitor*, June 16, 1998.

11. M. Traugott, P. Lavrakas, *The Voter's Guide to Election Polls* (New Jersey: Chatham House Publishers, Inc., 1996).

12. C. Turner *et al*, "Adolescent Sexual Behavior, Drug Use, and Violence: Increased Reporting with Computer Survey Technology," *Science*, May 8, 1998.

13. M. Wheeler, *Lies, Damn Lies, and Statistics: The Manipulation of Public Opinion in America* (New York: Liverright, 1976), p. 161.

24

Interpretation:
What Do the Numbers Tell Us?

24.1　Preface

How numbers are interpreted will, at best, depend on the judgment and capacity of the interpreters and, at worst, on the cleverness of the spin doctors entrusted with putting on them the best possible spin. It is often the case that "experts" see very different things in the same numbers and that different numbers concerning the same issue seem to be contradictory.

It is, I submit, most important to open up the issue of number interpretation to our students. A popular view has it that numbers contain within them one and only one meaning, the one "I" subscribe to—no ifs-ands-or-buts.
The following illustrations might be helpful for obtaining perspective on this dimension.

24.2　Math Tea Leaves

Case 1　Discrimination or Difference?

Concerned about charges of subtle patterns of bias at its executive levels, The United Federation of Worlds set up a Commission to investigate. During its hearings Lork from Mork pointed out that while Morkians are 30% of the Federation's work force at its lower levels, they make up only 1% of its executive staff. "Good faith recruitment efforts have been made," observed Lork, "but the statistics show subtle patterns of discrimination against Morkians." Tallia from Talos I disagreed. "The statistics show discrepancies," countered Tallia. "The subtle patterns of discrimination are your interpretation of the discrepancies

Who is right? We would have to agree with Tallia that the statistics show discrepancies. How the discrepancies are interpreted is another matter. Lork is offering one interpretation of the discrepancies that the statistics reveal.

What other explanations might account for the discrepancies?

If Case 1 strikes you as a less heated version of the discrimination battles in our nation, you are of course, right on target.

Case 2   How Successful Was the Patriot?

During the Gulf War television viewers were moved by scenes of Patriot missiles streaking across the sky to intercept Iraqui Scud missiles that had been launched against Israel and Saudi Arabia. The Patriot's success seemed to epitomize the success of a high tech, low causality military campaign.

Apart from very successful military public relations, how successful was the patriot as a military tool? Different statistics have been given and, as is almost always the case, it is important to look further than the statistics if a realistic picture is to emerge.

The Army originally stated that Patriots "intercepted" 45 of 47 incoming Scud missiles, and President Bush revised that to 41 of 42. What does this mean? Brigadier General Robert Drolet of the Army's Missile Command testified that "a Patriot and a Scud passed in the sky." There are other statistics of interest. Before Patriots were employed in Israel, 13 Scuds fell near Tel Aviv. There were no deaths, but 115 people were wounded and 2,698 apartments were damaged. After Patriots were employed in this region, 11 Scud attacks left 1 dead, 168 injured and 7,778 apartments damaged. (see [3] and [5]). This is explained by the fact that successful hits led to more deadly debris being sprayed over a larger area than otherwise would have been the case and that the Patriots tended to strike the bodies of Scuds, leaving their warheads armed and able to cause significantly damage on landing. But then it should also be kept in mind that the Patriot was not designed as an antimissile weapon, but to defend against fast-flying aircraft.

Case 3   Test Scores Rise: Education Miracle or Mirage?

In an era marked by bleak news on the education front with school systems finding themselves more and more driven to distraction, a June 1995 announcement that math and reading scores of elementary school students rose in all 32 New York City school districts that spring was greeted with pride and optimism that the City had turned an education corner. School officials attributed the rise to a new curriculum framework, improved guidelines for what children should be learning and when, and extra help for teachers in the City's 100 lowest performing schools.

But is there as much here as meets the eye? To what extent can an increase (or decrease, for that matter) in test scores be attributed to mastery of subject matter and to what extent can it be attributed to mastery of the measuring instrument—the exam itself? With standard exams that are continually being administered, grades tend to improve as teachers master the exam format and become adept at running practice sessions. (This is called teaching to the exam in those extreme cases where preoccupation with the exam overrides preoccupation with the subject itself.) Grades tend to decline when there is an abrupt change in the exam format which teachers and students have not had time to adopt to (such exams are often called unfair).

Case 4   There's Less to Baseball Statistics Than Meets the Bat?

In his book *The Last Yankee: The Turbulent Life of Billy Martin,* David Falkner [2] concludes that Martin was the best manager of his era, possibly of many eras. Falkner's judgment was strongly influenced by baseball statistics compiled by the Elias Sports Bureau and a formula which claims to show which managers' teams won more games than they were reasonable expected to win.

In his review of Falkner's book George F. Will [6] disputes Falkner's conclusion which, he argues, the rest of the book refutes. "In fact," Will notes:

The Last Yankee might usefully be made required reading for graduate students in the social sciences and all others who need to be immunized against the seduction of numbers . . . . There are limits—and Mr. Falkner's reporting shows that Elias passed them regarding Martin—to the ability to capture messy reality in tidy formulas.

Case 5   The Reagan Economic Boom: Blessing or Disaster?

Martin Anderson, former advisor to President Reagan and senior fellow at the Hoover Institution, employs statistics to support his view that the Reagan economic boom was the greatest ever [1].

Anderson's View

The two key measures that mark a depression or expansion are jobs and production. Let's look at the records that were set.

Creation of Jobs

From November 1982, when President Ronald Reagan's new economic program was beginning to take effect, to November 1989, 18.7 million new jobs were created. It was a world record: . . . The new jobs covered the entire spectrum of work, and more than half of them paid more than $20,000 a year. As total employment grew to 119.5 million, the rate of unemployment fell to slightly over 5 percent, the lowest level in 15 years.

Creation of Wealth

The amount of wealth produced during this seven year period was stupendous—some $30 trillion worth of goods and services. Again, it was a world record . . . . According to a recent study, net asset values—including stocks, bonds and real estate went up by more than $5 trillion between 1982 and 1989, an increase of roughly 50 percent . . . .

Income Tax Rates, Interest Rates and Inflation

Under President Reagan, top personal income tax rates were lowered dramatically from 70 percent to 28 percent. This policy change was the prime force behind the record breaking economic expansion . . . .

The Stock Market

Perhaps the key indicator of an economy's booms and busts is the stock market, the bottom line economic report card . . . . starting in late 1982, just as Reaganomics began to work, the stock market took off like a giant skyrocket. Since then, the Standard & Poor's index has soared, reaching a record high of 360, almost triple what it was in 1982.

There were other consequences of the expansion. Annual Federal spending on public housing and welfare, and on Social Security, Medicare and health all increased by billions of dollars. The poverty rate has fallen steadily since 1983.

When you add up the record of the Reagan years, and the first year of President Bush . . . the conclusion is clear, inescapable and stunning. We have just witnessed America's Great Expansion.

Leontief's View

In a reply, Nobel prize winning economist Wassily Leontief [4] concedes some of Anderson's statistics but goes on to look at a number of cost thorns in his statistical rose garden.

True, the long recovery from the deep depression that brought President Reagan to power carried this country to the high point of the usual cyclical wave characterized by a low rate of unemployment and a high gross national product. It is most likely that wholesale tax cuts inaugurated by Mr. Reagan have made the level of the G. N. P., as measured by the Government statisticians, several billion dollars higher than it would otherwise have been. But at what a cost.!

Drastic cuts in public spending (except for military purposes) left the physical infrastructure of this country in ruin. City streets and transportation facilities, water-supply and sewage systems, particularly in large metropolitan areas, are collapsing, the once glorious interstate highways are crumbling, and cramped airports are incapable of handling the rapidly increasing traffic. Despite the valiant effort of the underfinanced, underpowered Environmental Protection Agency, our lakes, rivers and forests are succumbing to deadly acid rain.

What is even worse, the intellectual, cultural and social infrastructure of the country has suffered even more during this greater-than-ever boom than its physical counterpart. Primary and secondary schooling have been so weakened that a whole generation of boys and girls can hardly read, write or count, while the soaring price of higher education makes it impossible for many young people to take advantage of it.

No wonder the competitiveness of the United States is rapidly declining; many of our high technology industries are losing one battle after another in the struggle for their share of the foreign and even their own domestic market. At the same time, the rich are getting richer, and the poor are getting homeless.

Let us hope that contrary to Mr. Anderson's expectations the "Reagan boom" will not continue in its present form for four or eight more years. If it does, the United States will find itself entering the 21st century as the richest country (in total value of stocks and bonds traded on the stock exchanges), but culturally and socially less advanced than other developed countries.

24.3   Food for Thought Questions

1. Continuing their discussion of what numbers show, Talia brought up the statistics concerning the Universe Games,. "Consider the Universe Games held every hundred years. For the last two thousand years 75% of the participants chosen by the trials have been Morkians. Does this mean that the trials were biased in favor of the Morkians?" "Certainly not," answered Lork, "they earned the right to be there in the trials that were held."

(a)   Do the numbers mean that the trials were biased in favor of Morkians? Explain.

(b)   Are there other "explanations" that might account for the numbers? Explain.

2. One thousand senior high school students in Ralph City took a college level math course, with 80% of them passing the statewide standard exam. The following year 3000 senior high school students in Ralph City took the course, with 50% of them passing the statewide standard exam. Do these figures gives us cause for pessimism or optimism concerning Ralph City's education system? How so?

3. "80 percent of the A grades in my math class were earned by Plutonians," commended Bob Levy to his colleague Amy Flores. "This means that they have an innate ability for math." Are there other interpretations of this figure? Are there other figures that might be relevant? Discuss.

4. Ellen Ames and Ann O'Neil, professors of economics at Huxley College, saw a grade of 50 on Professor Ames's last economics exam in different terms. "A grade of 50 is an F," noted Professor Ames. "But it's the highest grade in the class," replied Professor O'Neil, "and as such an F does not make sense." Discuss the significance of 50 as the highest grade on an exam, and as an F.

5. "80 percent of the crimes in this city were committed by Plutonians," remarked Oscar to his wife Janet. "They're a bad lot." Are there other interpretations of this figure? Are there other figures that might be relevant? Discuss.

6. "Blacks Prone to Dismissal By the U.S.: Their Discharge Rate Tops Other Workers," (K. De Wit, *The New York Times*, April 20, 1995, A19). Does this mean that blacks have been discriminated against when it comes to dismissal? Discuss.

7. "The proportion of black and Hispanic students in New York City public schools who read and do math at the level expected for their grade is far below that of white and Asian students, new test results show." C. Jones, "Test Scores Show Gaps by Ethnicity,' *The New York Times*, July 8, 1994, B1; A. B. Jeffries, "Schools Can't Give Students What They Don't Get at Home," *The New York Times*, July 20, 1994, A18.

    (a)   What are some possible explanations for these test results?

    (b)   What reforms do the explanations suggest?

8. Medicare, which provides health care for about 37 million Americans, is about to go broke, we have been told by Democrats and Republicans. Assertion: The Republicans intend to cut Medicare. Reply: Republicans do not intend to cut Medicare, just slow the rate of growth in spending. What do the figures suggest? Which figures are most meaningful and from what point of view? (See, for example, D. Rosenbaum, "The Medicare Brawl: Finger Pointing, Hyperbole and the Facts Behind Them," *The New York Times*, Oct. 1, 1995, 18.)

9. New York State education commissioner Richard Mills called the improved scores 'a success story across the board.' Is this the only interpretation that may be offered? Explain. Y. Zhao, "8th-Grade Math Test Scores Rise, But Most Fall Short of the Mark," *The New York Times*, Sept. 14, 2002; B1.

10. "New-home sales plunge, goods orders rise and jobless claims proliferate." How do these results add up? "Economic Data Point in Both Directions," *The New York Times*, Feb. 28, 2003; C3.

11. "Almost 40 percent of black men in their 20's in California are imprisoned or on probation or parole on any given day, a study released yesterday says." Do the numbers really imply discrimination? F. Butterfield, "Study Examines Race and Justice in California," *The New York Times*, Feb. 13, 1996; A12.

12. How so? G. Kolata, "Different Conclusion From the Same Study," *The New York Times*, April 9, 2002. F4.

13. Is the economy pulling itself out of a slump, or is it sinking deeper?" How do you interpret the numbers? D. Altman, "Data in Conflict: Why Economists Tend to Weep," *The New York Times*, July 11, 2003; C1

14. "Results, While Cheered, Drew Some Skepticism," How so? E. Gootman, "Math Scores Rise Sharply Across State," *The New York Times*, Oct. 22, 2003; B1

15. For the first half of 2009, "from New York to Los Angeles to Madison, Wis., major crimes, violent or not, are down between 7 percent and 22 percent over the same period last year. In Chicago, the number of homicides dropped 12 percent. In Charlotte, N.C., hard hit by the banking crisis, that total fell as astounding 38 percent."

    (a) What are the "mainstream explanations" of the cause of crime?

    (b) What policies do these "explanations" suggest?

    (c) What significance does the aforenoted crime statistics have for "mainstream explanations" of the cause of crime and the policies they suggest?

24.5   References of Interest?

1. M. Anderson, "The Reagan Boom—Greatest Ever," *The New York Times*, Jan. 17, 1990.

2. D. Falkner, *The Last Yankee: The Turbulent Life of Billy Martin* (New York: Simon & Schuster, 1991).

3. J. Jagger, "Why Patriot Didn't Work as Advertised," *The New York Times*, June 9, 1991.

4. W. Leontief, "We Can't Take More of this 'Reagan Boom,' *The New York Times*, Feb. 4, 1990.

5. E. Marshall, "Patriot's Scuds Busting Record is Challenged," *Science*, May 3, 1991.

6. G. Will, "Paranoid in Pinstripes," Review of [2], *The New York Times Book Review*, April 5, 1992, p. 17.

25

Why Do "Bad" Numbers (and Models, we Might Add) Matter?

25.1   Preface

We, as teachers of mathematics, are often asked by our students about the relevance of mathematics to them in terms of their quality of life.

There are many dimensions to quality of life, but one dimension that almost all of us would agree with is centered on decision making. While "good" numbers and "good" models (in the sense of being close to reality's mark) do not by themselves guarantee appropriate decision making, "bad" numbers and models (in the sense of being considerably off reality's mark) almost certainly guarantee in-appropriate decision making. This can be seen in both small and large scale scenarios.

Bottom-line Bob's misunderstanding about the nature of math modeling—"it's a no-brainer; 60,250 is larger than 51,000 . . . it's infallible; go with the Aleksa Company's solution since we want to maximize profit" (Introduction, p. 09—had disastrous consequences for the Austin Company when Bob's recommendation was implemented. P.M. (the practical man) found himself $400 poorer when he implemented S.P.'s recommendation based on the "bad" number, $\frac{1}{6}$, about the probability of an even number showing on a toss of P.M.'s die (Introduction, p. 09).

Large scale scenarios concerning inflation, presidential elections, and financial meltdown are considered in the three sections that follow.

25.2   Inflation

The Consumer Price Index (CPI) is the most generally accepted way of measuring inflation (ch. 10; sec. 10.4). Its value, based on the CPI model formulated by the Bureau of Labor Statistics (BLS), has an enormous inpact on the nation and on us individually (sec. 10.5). The monetary stakes are hugh and the question that inevitably follows is, how good/bad is the CPI value and model on which it rests (sec. 10.7)?

Under calmer circumstances this issue is left to the BLS, but in the 1990's the CPI took on a political dimension due to the balance-the-budget mantra that had arisen. This fascinating story is the subject of sec. 10.8.

25.3   Hayes vs. Tilden, Bush vs. Gore

Did questionable numbers determine the outcome of the 1876 presidential election? What were the consequences? N. Kleinfeld, "President Tilden? No, but Almost, in Another Vote That Dragged On," *The New York Times*, Nov. 12, 2000; 30. "Watch Yourself at Dinner, Dubya," *The Economist*, Nov. 25, 2000; 30.

In America's presidential election held in November 2000 George W. Bush defeated Albert Gore, Jr. to become President of the United States. At the end it came down to the state of Florida. Are the numbers and court decisions behind the numbers credible? "Supreme Court's Decision to Halt the Florida Recount," *The New York Times*, Dec. 10, 2000; 45; "How the Supreme Court Voted," *The New York Times*, Dec. 14, 2000; A33; A Lewis, "A Failure of Reason," *The New York Times*, Dec. 16, 2000; 16; L. Greenhouse, "Collision With Politics Risks Court's Legal Credibility," *The New York Times*, Dec. 11, 2000; A1; D. Schorr, "The Supreme Fix Was In," *The Christian Science Monitor*, Dec. 15, 2000; 11; R. Dworkin, "A Badly Flawed Election," *The New York Review of Books*, Jan. 11, 2000; 53-55; "Opening a Gavel of Worms," *The Economist*, Dec. 16, 2000; 30, 33.

Hayes or Tilden, Bush or Gore. It, needless-to-say, goes beyond names. The key question is, what difference did it make to the country in terms of their outlook, policies, and accomplishments.

25.4   Value at Risk (VaR) Models

VaR models achieved great popularity in the 1990's. Their great appeal is that they yield risk as a valid conclusion in terms of a dollar figure. Alas, the risk factor the VaR's did not take into account in their **assumptions** is the risk of a financial meltdown (which came to pass in 2008-2009). See [8] and ch. 2, sec. 2.9, question 9.

25.5   Pennywise and Pound Foolish

What can we do to support the research and data collection needed to make the CPI model, among others, as realistic as possible?

The problem is to obtain sufficient funding to do a proper job of number crunching, which has become an annual battle with Congress. Murray Weidenbaum, chief economic advisor to President Reagan, was prompted to observe:

> The real costs resulting from providing inadequate funding for one of the most useful parts of the bureaucracy—the data gatherers—are truly awesome. An inaccurately high report of inflation can trigger an avoidable policy of monetary restraint followed by needless declines in capital formation, production, and employment. Bad information on productivity can generate investor decisions out of sync with real trends in the marketplace.
>
> Underreporting exports can produce news of "record" trade deficits followed by adoption of tougher protectionist policies, which could undermine the growth prospects for the US as well as those nations it trades with. [10]

25.6   Food for Thought Questions

1. What impact did "bad" numbers have on the Energy Crisis of the 1970's? See [1] and [6].

2. Describe two other situations where "bad" numbers/models had an impact on decision making.

25.7   References of Interest?

1. J. Duncan and A. Gross, *Statistics for the 21st Century* (Chicago; Irwin, 1995).

2. "Damned Lies: Economic Statistics Are in a Bad Way," (Editorial) *The Economist*, Nov. 23, 1996.

3. E. Ehrlich, "The Downside of Bad Data," *Challenge*, March/April 1997; 13-35.

4. M. Gardner, "The Power of Statistics to Affect Lives—Even When They're Wrong," *The Christian Science Monitor*, May 2, 1996; 12.

5. H. Gleckman, "On Congress' Hit List: Crucial Business Data," *Business Week*, Sept. 13, 1999; 40.

6. E.C. Ladd, "How Bad Can Numbers Be? Remember the Energy Crisis," *The Christian Science Monitor*, June 2, 1995; 19.

7. M. Mandel, "The Real Truth About the Economy," *Business Week*, Nov. 7, 1994.

8. J. Nocera "Risk Management," *New York Times Magazine*, Jan. 04, 2009; 24-33, 46, 50, 51.

9. S. Strong, "The Link Between Quality of Data and Quality of Life," *The Christian Science Monitor*, Jan. 30, 1997; 19.

10. M. Weidenbaum, "Fund the Number Crunchers," *The Christian Science Monitor*, Sept. 16, 1999; 9.

26

Comments on Selected
Food for Thought Questions: 3

26.1 Chapter 22 (p. 416)

1. (a) No. The data, simply interpreted, might give us a measure of the "efficiency" or "productivity" of each instructor for each class, but this does not translate to a measure of the over-all "efficiency" or "productivity" of the department.

   (b) No. This game of academic musical chairs will not change the department's over-all tuition revenue or costs, on which financial efficiency depends.

2. No; the number of years driving accident free says nothing about the number of miles driven, which is the more telling figure.

3. To help get a better grip on this let us put some numbers to the ingredients. Let us suppose that the English department of 10 faculty, to continue with our example, has an average (annual) salary of $70,000 and that each faculty member is required to teach 7 courses (or sections) during the academic year. Then the average cost per course is $10,000. If each student is charged $1,000 tuition for a course, then 10 students are needed

to meet the average cost for the course and 15 are needed to satisfy criterion RC-2.

(a)   The salary of each department member is needed so that the average salary for that department can be determined.

(b)   As to merits, RC-2 is an improvement over RC-1 in that the run/cancel decision for a course is not dependent on the instructor assigned to the course but on the number of students enrolled in it, which makes more sense. RC-2 is clearly easier to implement than RC-1.

As to disadvantages, if the department consisted mostly of relatively high paid senior faculty with an average salary of $105,000, for example, an enrollment of 20 in a course would be needed to satisfy RC-2. If the department consisted mostly of recently hired junior faculty with an average salary of $35,000, let us assume, an enrollment of 10 in a course would be needed to satisfy RC-2. The implementation of RC-2 might vary considerably from department to department depending on the mix of the department's junior and senior faculty.

This does not make good academic sense. Departments are not unconnected units to be considered in isolation, but are intended to service a larger whole—the university itself. The course run/cancel criteria should reflect the needs of the university as a whole and the shortcoming of RC-2 is that is it not designed to do this.

26.2   Chapter 23 (p. 440)

1   (a)   No; the homeowners listing of Bell City does not describe the public at large.

(b)   The appropriate target audience is the public at large.

(c)   No; the sample drawn is not representative of the public at large.

2   (a)   Agree

(b)   Only one option is offered—take military action within six months.

(c-i)   There are, of course, many other possible options that might have been made available.

(c-ii)   For example, how important do you think it is for the U.S. to work with the U.N. in resolving its differences with Iraq; should military action against Iraq be taken by the U.S.? An important response option, not included here, is Need More Information.

(c-iii)   I don't see how you can suitably modify the wording of the question posed to make it less of a leading question. To make it more "neutral" it would have to be dropped and replaced by more neutral alternatives or kept with other options included.

3.   (i)   Yes.

ii(a)   "Take more money out of your pocket" in (a) makes the question biased; it is clear from this phrase that the Center for Critical Thinking does not favor raising taxes and is leading potential respondents to the question in the same direction.

> The phrase "defense capability reduced by budget cuts" makes (b) biased.

> The word "crushing" makes (c) biased.

> The word "squander" makes (d) biased.

ii(b)   Bias can be subdued by dropping the inflammatory phrases or words. For (a) we have: would you vote for a presidential candidate who was willing to raise taxes?

4. (a)   No; Yes or No is very restrictive. It does not allow for the numerous shades of gray that a potential respondent to the question might find more compatible with his views.

    (b)   Not sure; need more information.

6. No. He only heard from people who, for one reason or another, decided to call. The rest of the voters could possibly turn the tide.

7. They made **assumptions** on how the undecided would vote and they were wrong. They also **assumed** that public opinion would remain relatively stable from the time of their last polls in September to election day in November, which proved to be wrong.

20. 'The substitution of just one word, or the order in which questions are asked, can have a profound effect on the results,' said Mark A. Schulman, the president of the American Association for Public Opinion Research. Asking someone, for instance, whether they think Mr. Hussein 'helped the terrorists' in the Sept. 11 attacks, as the Pew Research did last October [2002], will often yield more positive responses than would questions that ask if Mr. Hussein was 'personally involved' in the attacks, as a Time/CNN poll did at roughly the same time.

26.3   Chapter 24 (p. 457)

1. (a)   The statistics, by themselves, show discrepancies. Whether or not the discrepancies translate to bias is another matter.

    (b)   The Morkians may possibly come from a culture that places a high premium on games and competition and this emphasis may be showing up in the Universe Games.

3. First of all, the 80 percent figure is just that: it does not, by itself, determine an interpretation. As to Bob Levy's interpretation, it's quite a stretch to go from the 80 percent figure to the conclusion that Plutonians have an inate ability for math. Another possible interpretation is that the Plutonians in the class, at least those who got A's, are unusually well-prepared and hard working.

4. Both interpretations as well as a number of others are feasible. Which of the two interpretations offered here is the more credible depends on how a number of background factors are taken into account. These factors include level of difficulty of the exam (tough vs. easy vs. various in-betweens), and the commitment of the students to the course (lackadaisical vs. serious vs. various in-betweens).

   The important point to keep in mind, I reaffirm to my students, is that the number 50, by itself, does not settle the issue of interpretation.

27

Return to The Question of Math Insight, Myth, in a Sense a Bit of Both, or Nonsense

27.1   Preface

May I invite you to address the views expressed about mathematics in the Introduction from the point of view of the afore question. I suggest that you only turn to my comments after you have addressed the question yourself.

27.2   A presentation supported by figures is more credible than one that is not.

Myth: the illustrative examples discussed in ch. 21 along with the suggested food for thought questions show that the reliability of many of the numbers paraded before us are seriously open to question.

27.3   Numbers are neutral. They are not affected by cultural, economic and political differences that influence people and therein lies in their strength. 1 + 1 = 2, for example, is universal for all peoples, countries, and political systems.

Myth: 1 + 1 = 2 in mathematics, irrespective of cultural, economic, and political differences in cultures, but when it comes to a real world situation such as counting the number of successes in testing a missle defense system, for example, the first 1 describing success on the first test, and the second 1, describing success on the second test, adds to 2 successes for testing the missle defense system provided that the two claimed successes is trustworthy. See, for example Case 3 (Top of the Line Deception), ch. 21.

27.4   Numbers give weight to a view through the sense of *precision* they communicate, thereby advancing it to a plane which commands recognition, respect, and acceptance.

In a sense a bit of both: " . . . thereby advancing it to a plane which commands recognition, respect and acceptance" is the gold we are seeking, but then there is the gold look alike called fools-gold (pyrite). For numbers fools-gold translates to numbers that are not trustworthy (ch. 21) or not relevant (ch. 22).

27.5   Mathematical proof is the most reliable means for objectively establishing truth.

Nonsense: this statement is a tenet of the Uncle George School of Thought (ch. 4; sec. 4.3). Replace truth by validity and the statement reflects a math insight.

27.6   The *precision* of mathematical methods guarantees unassailable conclusions which serve as pillars of stability and strength in a world besieged by foggy thinking, prejudice, and rampant special interests.

In a sense a bit of both: it depends on the use of the word *precision*. Is it *precision* in the sense of realism or *precision* in the sense of validity?

The coloring " . . . guarantees unassailable conclusions . . . in a world besieged by foggy thinking, prejudice . . ." might be interpreted as meaning that *precision* in the sense of realism is what is intended; if so, the statement is a math myth. If *precision* in the sense of validity is what is intended, the statement is a math insight.

27.7   Mathematically derived conclusions are indisputable because they are based on deductive logic, which is untainted by bias and ideology.

In a sense a bit of both; just as *precision* is the key word in 27.6, here it is indisputable. Indisputable in the sense of validity, or indisputable in the sense of realism? " . . . because they are based on deductive logic . . ." means the former, strictly speaking, which makes the statement a math insight.

If indisputable is understood in the sense of realism—which the coloring " . . . untainted by bias and ideology." might suggest to some—then the statement is a math myth.

27.8   Considering the availability of powerful computers, in applications of mathematics the hypothesis of theorems warrants minor consideration.

Nonsense: availability or non-availability of computers, powerful or not, is irrelevant to the issue of the level of attention that must be given to the hypothesis of a theorem. The hypothesis of a theorem is the foundation stone on which the valid conclusion of the theorem rests. It must be given highest priority, period.

Index

F